FIBRE OPTICS

FIBRE OPTICS

By D. A. Hill

BUSINESS BOOKS
COMMUNICA - EUROPA

First published 1977

ISBN 0 220 66333 5

This book has been set 11 on 12 pt Press Roman.
Printed and bound in England by
Redwood Burn Limited, Trowbridge and Esher,
for the publishers, Business Books Limited,
24 Highbury Crescent, London N5

CONTENTS

PREFACE

For many years fibre optics was a solution without a problem. Since the 1950s there has been a rapid advance in applying the solution to a whole range of problems and there have also been developments in the components upon which fibre optics, along with other technologies, depends.

The question is often asked 'What is the main use of fibre optics?'. The uses are by now so extensive as to render it impossible to give a short answer to that question.

Kenilworth
June 1977   DENYS HILL

ACKNOWLEDGEMENTS

Writing a book makes one very conscious of the value of friends.
Heinz Brachvogel has made available not only his collection of
unpublished papers on fibre optics but also his extensive knowledge
of optics. Werner Scholly has read my manuscript and placed at my
disposal his detailed knowledge of the applications of fibre optics.
Ted Rockley has, at every stage in the preparation of my manuscript,
allowed me to draw upon the experience which he, as the author of
several books, has accumulated. Mrs Vera Collett has demonstrated
her ability to transcribe my hand-writing into typescript – no small
achievement!

On a more formal note I wish to acknowledge the help of
Electronics Today International for permission to quote from
Professor W.A. Gambling's article, 'Laser Communications'
(Volume 5, No 7).

None of the above should be held responsible for my deficiencies
or errors.

INTRODUCTION

The two words, *fibre optics,* conjure up several different concepts
and some misunderstandings. To some, the words denote those
weeping willow-like bundles of greyish white filaments, fixed at one
end and gently curving along their lengths, each filament freely
flexing, in response to air movements, inside the room where the
lamp, of which these filaments are the decorative component, is
placed. The filaments transmit both white and coloured light.

Viewers of popular broadcast scientific television programmes
may associate fibre optics with a particular type of fibre-optic
instrument — usually a fibrescope. Although they have industrial uses,
the members of the flexible fibrescope family have a significant role
as a diagnostic aid in medical science because they permit the tortuous
internal passages of the human body to be visually examined without
surgery.

Different industries associate fibre-optics technology with over-
coming or lessening problems of quality assurance, inspection,
machine and process control. By common usage, fibre optics has
two meanings. The two words describe a technology as well as its
principal component — optical fibres. The technology embraces not
only optical fibres but light sources, instruments incorporating optical
fibres and the uses of optical fibres in their assembled state.
Throughout the book, care has been taken to distinguish the component from the technology.

The youthfulness of the technology may be the principal reason
why there are misunderstandings in fibre-optic nomenclature. The
filaments in the decorative lamp are optical fibres, to use the precise
description, not fibre optics. Optical fibres are sometimes called
light pipes in the belief that they are flexible, small-diameter
capillary tubes. They are solid, each being composed of two different
types of translucent material, and transmitting light according to the
principle of total internal reflection.

A fibre optic is an expression occasionally used to describe a bundle of those fine optical fibres without indicating whether they are randomly gathered or coherently orientated, fibre by fibre. There is imprecision in respect of the terminology used to describe the two parts of optical fibres and the protective coverings which enclose many of them. Each optical fibre, irrespective of the material from which it is made, natural or synthetic, has a central core and usually only one outer cladding layer. The cladding layer is sometimes known as the *sheath* but the protective outer covering of a bundle of fibres is also termed a sheath.

Cladding has been retained for the outer skin of optical fibres and *jacket* is used for the protective covering. *Image bundle* is used for coherently assembled fibres, and *light conduit* for randomly assembled ones in a rigid jacket. Since all fibres conduct light in some way, the words *light conductor* have been avoided for randomly assembled fibres in a flexible jacket. Two different expressions are used. The expressions vary with the fibre-optic application.

Bundles of fibre optics used for transmitting light in one direction, either independently or connected to a fibre-optic viewing instrument, are termed *lightguides*. In the opto-electronic context, where scanning and process control fibre-optic instruments are used, the word *probe* is used. Probes transmit light by a coaxial arrangement of optical fibres in both directions. This ability to transmit light in either direction along straight or curved paths, even narrow tortuous ones, is the foundation of fibre-optics technology and a useful component for other technologies.

Optical fibres have four different groups of spatial applications:

1   They are used to transmit or transport low-energy light energy from a *confined* space to an *unrestricted space*. The decorative lamp takes light from a lamp fixed in a small base to the unrestricted ends of the fibres. Lightguides inside vehicle lamps convey, to the relatively unrestricted space of the car interior, simple analogue information as to whether head, side and rear lamps are functioning when switched on (Figure I.1).

2   Optical fibres, often inside a fibre-optic viewing instrument, are used to transmit visible light from an *unrestricted space* to a *confined* one, for the examination of small, otherwise inaccessible, objects. The later word *object*, is used in the optical sense of denoting the thing which is viewed, be it an area or a point.

a   A bundle of fine optical fibres or a thicker mono-fibre, connected at the unrestricted end to a light source and eased into a small straight hole which has to be examined for unwanted features, e.g. burrs, constitutes the first sub-division of this second spatial fibre-optical application group (Figure I.2).

b   Other lightguide configurations are used with non-fibre-optic instruments, such as microscopes and profile projectors, to

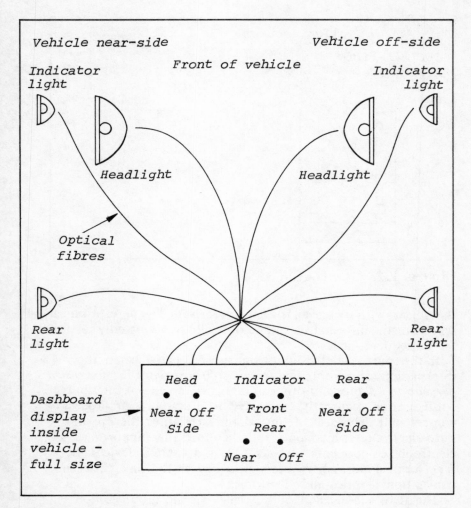

Figure I.1 Vehicle light monitoring

take light into the confirmed spaces between the object lens
and the specimen. Yet other assembled configurations
distribute high-quality light over a specimen. Optical fibres
can be assembled in so many different ways. They are therefore also used inside computer card and tape readers.

c   Inside both rigid and flexible fibre-optic viewing instruments
optical fibres function alongside other optical components such
as lenses, prisms and mirrors. Rigid endoscopes use lenses for
image transmission whereas flexible fibrescopes transmit their
information, in the parlance which is being ever more widely
used, through coherent optical fibres.

d   Single filaments of self-focusing uncladded fibre (see page 32)
are being used as an optical system for heart and brain research.

In all three of the above sub-divisions the light, after returning from

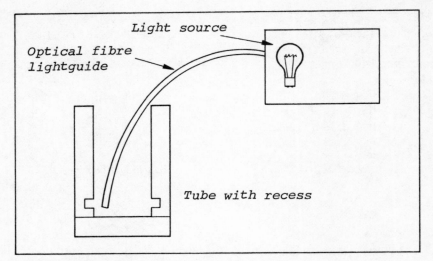

Figure I.2 Lightguide in narrow tube

the object, with its image, is visibly perceived. The perception can be on-line or the information can be statically or animatedly recorded for subsequent retrieval.

In the fourth sub-division of this second spatial group, the information can be obtained from a *confined space* for *non-visual perception.* A probe or other suitably assembled optical fibre configuration is directly connected to a scanning or process control unit. This use of optical fibres requires the use of opto-electronic components and therefore the supporting electronic components such as TTL and C-MOS devices. In fact, optical fibres have been described as the link between conventional optics and electronics.

The fifth sub-division of this large second group is that where *short lengths* of coherent fibres are *fused* together. They are used as lenses to collect light in an unrestricted space for improved transmission to a confined space. A fibre-optic faceplate which is used as an image intensifier in front of a television camera is an example.

3   The last main group of fibre-optic applications, spatially clas-sified, is that of transmitting invisible light from one *unrestricted* space to *another unrestricted space* in the growing field of fibre-optic telecommunications. The light may be emitted by a laser. Audio-visual information is converted to light valves for transmission and then re-converted back to its original form. Occasionally optical fibres transmit between *two restricted* spaces.

In all the main groups of applications, fibre optics either comple-ment, replace or pioneer. Light can be taken round irregularly curved paths, with relative ease, but the multitudinous combinations of conventional lenses, prisms and mirrors could not do that. Scientific instruments cannot enter long, small-diameter holes nor are electric lamps capable of emitting a useful amount of light. Fibre optics can. Microscopes cannot view objects when a curved access has to be negotiated. Copper wire cannot transmit electrically inert light. Fibre optics can.

When incorporated in a fibre-optic viewing instrument, alongside conventional optical components, optical fibres open up a new, narrower and longer group of applications for those optical components. A fibrescope may be over 4 m long, yet have a diameter of only 10 mm. Self-focusing fibres have a diameter of 150 μ and act as an optical instrument.

Interfaced with opto-electronic emitters and receivers, optical-fibre probes enable even smaller targets to be detected at electronic speeds. Opto-electronics are variously termed electro-optics and photo-optics.

The practical applications of fibre optics are already extensive for a youthful technology. Drivers on motorways receive information from optical fibres arranged in raster or matrix form and mounted on the central reservations. Breakfast cereal eaters have unwanted burnt flakes rejected, before they reach the packet, by a fibre-optic process control unit. Domestic electric lamps are automatically assembled with the aid of a fibre-optic scanning unit. Civil, electrical, mechanical, automotive and electronics engineers use fibre optics to examine for defects. Carpet patterns are read by fibre optics. Sportsmen have their performances displayed as they happen. Photographers gain better illumination of their subjects. Watches are assembled more quickly. Microscopists view more clearly. The list is long but it is augmented in the appropriate passages of the book and more fully described.

The material is presented so as to answer four questions about fibre-optics technology:

1 How does it work? (The theory.)
2 What does it comprise? (The components.)
3 What is used? (Fibre-optic instruments.)
4 How is it used? (Fibre optics for . . . .)

With few exceptions, the material is presented in non-mathematical form. Where necessary to gain an understanding of fibre optics, simple mathematics are introduced. The more detailed mathematical treatment of the subject of light and electronics is available in the standard text-books.

The first chapter, 'Fibre optics and light', relates the theoretical data to fibre optics, in narrative form. The second chapter, 'Basic fibre-optic components', describes optical fibres in detail and the other components with which they are associated. The third chapter, 'Fibre-optic instruments', groups the components together. The final chapter groups the instruments in the combinations in which they have been used to overcome the theoretical and practical problems of real-life situations. In all the chapters some objectivity has been introduced to avoid affording fibre optics an insular treatment which it could not expect in the work situation where it has to compete with alternative technologies. The sub-title 'Limitations' appears in several places and reference is made to other technologies, devices and methods of working.

The organisation of the fibre-optics industry is reflected in three of the chapter headings. Although not rigidly demarcated, the industry can be divided into optical-fibre manufacturers and fibre-optic equipment manufacturers, the latter also discharging the role of applications engineers. Some components are identical to those used by makers of non-fibre-optic electronic and optical equipment.

Optical-fibre manufacturers are situated in the traditional optical-glass manufacturing areas in Europe, North America and Asia. Synthetic fibres are made by some manufacturers who supply synthetic yarns for the knitwear industry. Some of these fibre manufacturers assemble optical fibres but several confine themselves to supplying fibres to the manufacturers of fibre-optic equipment.

The manufacturing facilities are not duplicated in each continent. Although the scope of fibre optics is extensive, with the exception of telecommunications, labour costs are higher than material costs in the finished products. Some types of fibre are therefore available from only one source because the market is insufficient to warrant a second source of supply. Fibre manufacture requires complicated manufacturing processes so 'economies of scale' is an economic concept very relevant to the organisation of the fibre-optics industry.

To the fibre-optic equipment manufacturer optical fibres are only one of several components. However, their optical affinities have caused the majority of them to be located in optical or precision engineering areas where personnel are accustomed to working to very close limits on small components. Fibre-optic telecommunications manufacturers, who are a specialist group of fibre-optic equipment manufacturers, have very close connections with 'conventional' telecommunications and electronics manufacturers. Many of them are part of the same firms (see Appendix 1).

Over the years, even before many firms commenced operations, there have been many discoveries and historical events that have laid a foundation for fibre optics. Many of them long preceded the discovery of optical fibres. These events and discoveries are listed below, not for geographical or historical reasons, but first to demonstrate that fibre-optics is broadly based, still evolving and closely related to other technologies. Thus, fears about using a novel technology with an avowedly meteoric growth can be dispelled. Secondly, the list provides an opportunity for the introduction, by way of annotation, of the many principles and components that fibre optics embrace.

The history has three constituent elements — the discovery of scientific laws, the discovery and evaluation of materials, and the manufacture of components and scientific instruments.

Chronologically, glass as a material, albeit not an optical one, preceded the discovery of scientific laws. Early Egyptian, Phoenician and Babylonian glass was rarely translucent. Transparent glass was developed in the centuries spanned by medieval history, not ancient. The development of optical glass followed the manufacture of the elegant Venetian glassware of the High Renaissance. The men of Jena, by making optical glass in the nineteenth century, made a significant contribution to the provision of raw material for optical fibres.

Discoveries about the laws of light are distributed throughout the centuries. Heron of Alexandria, Snell, Huygens, Herschel, Newton, Ritter and, in this century, Planck and Einstein, can all lay claim to significant contributions to optical theory.

Men like Galileo produced early scientific instruments which not only gave practical expression to theoretical laws, but introduced changes into the methodology of scientific investigation. It no longer needed to be philosophical or theological.

Discoveries in electronics, which are incorporated in fibre-optic scanning and process control units, have been made in this and the preceding century.

The chronological method of presenting historical data, whilst economical in terms of space has, for the purposes of this introduction to a book on fibre optics, one big disadvantage. Technological evolution cannot always be dated so the list does not fully portray the growth in fibre-optics since the 1950s even if it shows the earlier discoveries of the 1920s.

The following are some of the events and discoveries relevant to fibre optics:

BC 1546-1316   The El-Amarna glass factory in Egypt made small coloured ointment jars by casting or winding drawn glass rods round a sand core and then re-heating. The glass was not very translucent due to impurities in the basic materials. Cobalt and manganese were used as colouring agents.

c 200 AD   Heron of Alexandria discovered reflection. Total internal reflection is the principle by which light is transmitted by optical fibres.

c 1025   Alhazen discussed the magnifying effect of a simple lens. Lenses have many uses in fibre-optic instruments.

1590   Janssen invented a 'microscope'.

1604   Johannes Kepler applied the laws of rectilinear propagation to photometry — the measurement of light. Process-control units employ this principle.

1609   Galileo made a refractory telescope.

1621   Willebrod Snell discovered that when light travels from one transparent medium to another its direction is changed — refraction.

1669   Brandt discovered phosphorous compounds. Similar compounds are used in optical fibres for telecommunications to reduce attenuation.

1670   Bartholinus discovered that when a beam of light passes through certain crystals, e.g. calcite, each beam is split into two. Beam splitting is used with some low-power-magnification endoscopes.

1676   Romer discovered that light has a finite velocity.

1678   Christian Huygens discovered that light travels in waves. The wave theory of light affects the choice of optical-fibre materials and light sources.

1729   James Bradley discovered aberrations in light. The designer of the fibre-optic viewing instruments has to overcome these.

1758   John Dolland invented the achromatic lens (see 1729 above).

1774   Scheele isolated chlorine which is used with phosphorous and silicon in optical fibres for telecommunications.

1774   Joseph Priestley separated oxygen. It has applications similar to chlorine.

1790   Guinand discovered a stirring process for improving the homogeneity of molton glass.

1791   Luigi Galvani discovered the principle of the battery. Batteries are used in some light sources.

1791   Gregor discovered titanium. In its metallic form it is sometimes used to protect fibre-optic viewing instruments against heat (see 1928).

1800   Sir William Herschel discovered the infra-red part of the spectrum. Infra-red light is used in many fibre-optic scanning units.

1801   J. W. Ritter discovered the ultra-violet part of the spectrum.

1803 J.J. Berzelius discovered cerium, which is being used in experiments to improve the radiation resistance of optical fibres.

1817 J. J. Berzelius discovered selenium (see 1873).

1839 A. E. Becquerel observed the photo-voltaic effect. This is used in fibre-optic scanning and process control units.

1839 William Henry Fox Talbot invented the calotype photographic process which, together with the process invented in 1850, laid the foundations of modern photography. Photographic cameras are used with endoscopes and fibrescopes — fibre-optic viewing instruments.

1846 Carl Zeiss started manufacturing high-grade lenses, compound microscopes and cameras at Jena in collaboration with B. Schott and Ernst Abbe.

1851 Chance Brothers, with the aid of Bontemps, exhibited optical glass at the Great Exhibition in the Crystal Palace, London.

1860 Sir Joseph Swan made the first electric lamps. Thomas Edison subsequently developed them.

1870 John Tyndall demonstrated the basic principle of fibre-optics to the Royal Society in London. A stream of water flowing from an illuminated vessel, filled with water, conducted light.

1873 Willoughby Smith discovered photoconductivity in selenium.

1875 Lecoq de Boisbaudran discovered gallium, compounds of which are used in fibre-optic scanning units and telecommunications lasers.

c 1880 Robert Bunsen discovered caesium, which is used in television camera tubes.

1886 Winkler discovered germanium, compounds of which are used as an alternative to gallium in scanning units.

1895 Julius Elster and Hans F. Geitel invented the photo-voltaic cell.

1900 Max Planck established the quantum theory, in which light is one form of radiant energy and consists of discrete quantities of energy called photons.

1905   Albert Einstein explained the photo-voltaic effect.

1909   Leo Baekland discovered a viscose product as a result of a reaction between phenol and formaldehyde. This discovery of polymerisation was the forerunner of the materials from which synthetic optical fibres are extruded.

1915   G. W. Morey reduced glass melting time from three days to one day.

1917   Einstein predicted stimulated emission of light. Lasers are being used in fibre-optic telecommunications.

1919   A. O. Rankine described voice transmission by light — the principle of fibre-optic telecommunications.

1920   O. Schiever described experiments in dielectric wave guides — the forrunners of optical fibres.

1924   Louis de Broglie suggested that not only the energy of the electron magnetic spectrum shows a dual wave-particle nature but that all particles have a wave nature (see 1678 above).

1924   John Logie Baird obtained electronic images on a television screen. Television cameras and monitors are used, in a complementary role, with endoscopes and fibrescopes.

1926   Baird patented an invention for transmitting images through bundles of small diameter glass rods. This principle, much refined, is used in fibrescopes.

1928   William A. Knoll made the first titanium metal.

1932   Harold Urey discovered deuterium (heavy water), which is used in some ultra-violet light sources.

1938   Vladimir Zworykin invented the iconoscope, which improved television scanning and therefore image definition. The orthicon, plumbicon and vidicon have superseded the iconoscope.

1939-45   The use of radar during the Second World War increased the use of semiconductor electronic devices to such an extent that, at the end of the war, alternative non-military uses had to be found for these devices. They are used in fibre-optic scanning and process control units.

1948   Schockley discovered the much cheaper junction transistor.

1954 A. C. S. Van Heel suggested a method of transporting images without aberration. He also investigated, as did O'Brien in the USA, the problems of optical insulation. Both pieces of work were important for optical fibres.

1954 H. H. Hopkins and N. S. Kapany published an article, 'A flexible fibrescope using static scanning'.

1957 Vasco Ronchi defined an image as a recognisable non-uniformity of light.

1960 T. W. Maiman constructed the first operating laser.

1966 Kao and Hockham suggested a method for sending light over long distances — fibre-optic telecommunications.

1967 The first commercial electronic chip (C-MOS) was shown at the IEEE show in New York. These are used in some scanning units.

1970 Corning produced the first long optical fibre for telecommunications. It had an attenuation of 20 dB/km (32 dB/mile).

1975 A BBC colour television programme was sent through 1.24 km (2 miles) of optical fibre before being broadcast.

1975 W. A. Gambling, at Southampton, produced phosphosilicate optical fibres to overcome some intrinsic fibre-optic telecommunications problems. The attenuation was 2 dB/km (3.2 dB/mile).

The above list has introduced the three separate roles for fibre-optics — illumination for viewing, non-visual sensing and telecommunications. The next chapter shows how light, in its many forms, is the basis of all three functions.

One

FIBRE OPTICS AND LIGHT

1.1 Résumé

The theoretical aspects of light have been described in both narrative and mathematical form by a number of authors, for different purposes. The literature on light originated in the ancient world with the Greek philosophers. The purpose of this chapter is not to give a fuller history of the development of this branch of scientific theory than has already been given in the Introduction, but to describe those aspects of the subject of light, an understanding of which is a prerequisite for understanding fibre optics.

Since there is a considerable amount of good general literature on the subject of light, some of the material in this chapter is presented in summary form only. Teachers and students of physics will recognise nodal points where their knowledge of light is particularly relevant to fibre optics. Users and potential users of fibre optics will have an awareness of the theoretical considerations encountered in the practical applications described in the final chapter of the book. In order to avoid divorcing practical applications from the theoretical aspects of fibre optics, a limited number of examples are introduced in this chapter to illustrate how theory becomes manifest in practice These theoretical descriptions constitute the anatomy into which fibre-optic components, like organs of the body, can be fitted, so as to discharge, not physiological, but the four basic fibre-optic functions.

1.2 The definition of light

The debate concerning the nature of light, often referred to as physical optics, to distinguish it from geometrical optics — the way in which light is redirected — is a long-standing one. It began with the Greek philosophers and still continues.

The two basic theories or groups of theories, which have current credence, are the wave theory and the quantum theory. Optics can be largely explained in terms of wave theory but there are difficulties when relating it to those electronic devices which are used in fibre-optic scanning and process control units — hence the quantum theory. Attempts have been made to reconcile the two theories, not because fibre optics bridges electronics and optics, but because science requires that the inexplicable be explained.

There are variations on the wave theory. The analogy of waves on the surface of water is the starting point for explaining the theory. Figure 1.1 shows a view from above a series of waves. They can be waves of water reduced in size and straightened, or light waves, highly magnified, moving in only one plane. The thick lines represent the crests and the thin ones the troughs of the waves. Each crest or trough preserves its own individuality as it moves along. The distance between the crests is one wavelength — the distance can be measured. The distance between the crests of light waves is very small indeed, so small as to be measured in nanometres (nm): 1 nm = 0.000000001 m. Alternatively, light is measured in Ångstrom units (1 Å = 0.1 nm). Anders Jonas Ångstrom (1814-74) was a Swedish physicist who studied heat magnetism and spectroscopy. Sometimes microns are used.

Light waves do not always operate in the same plane. As they move forward they rotate irregularly about an axis unless they are polarised.

White light comprises several different wavelengths. The type of light source affects the energy level of each wavelength so that there are, in effect, different types of white light. Each individual colour, and the wavelengths invisible to the human eye have their own bands of wavelengths (see Figure 1.2). Ultra-violet has the shortest wavelengths and infra-red the longest (see Table 1.1). The materials from which optical fibres are made do not transmit all wavelengths of light with equal efficiency.

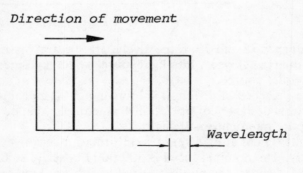

Figure 1.1   Enlarged representation of waves

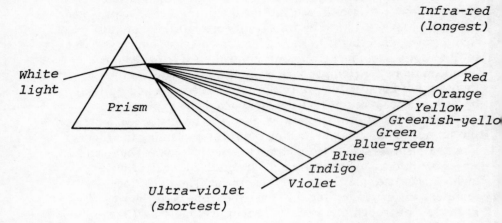

Figure 1.2 Dispersion of light

TABLE 1.1

Colour	Nominal wavelength, nm
Ultra-violet	170
Violet	400
Blue	480
Blue-green	520
Green	540
Greenish-yellow	560
Yellow	580
Orange	620
Red	660
Dark red	700
Infra-red	900

On pages 78 and 79, respectively, are spectral distribution diagrams for a quartz halogen white-light source and a high-pressure-mercury ultra-violet source.

The presence of different colours in light is a problem which confronts designers of optical instruments. Lasers, however, emit very pure monochromatic light.

Light travels at high speed so information is conveyed very quickly. The nominal speed is 300,000 km/sec (186,000 miles/sec). Table 1.2 shows the number of pulses for the three broad groups of light — ultra-violet, visible and infra-red, with which fibre optics is concerned.

Since light used with fibre optics is only one of several ways of obtaining information, some of them competitive but many mutually exclusive or complementary, Figure 1.3 has been included. Development work for transmitting radar through optical fibres is being undertaken to improve waveguide design.

The second group of light theories are the particle ones. There were difficulties for wave theories when electromagnetism had to be explained. About 1839 it was discovered that light could cause atoms to emit electrons. Later Einstein suggested that the energy of a beam of light is not spread uniformly over the whole beam but is concentrated in certain regions. He called these photons. These photons, he suggested, are propagated like particles. If a photon can be envisaged as a packet of electromagnetic waves, the two theories can be partially reconciled.

The photon or quantum is the basic unit of energy used in nuclear physics. In many respects it is a sub-atomic particle. The shorter the wavelength, the greater the energy of a photon. An X-ray has greater energy than a visible-light photon. Ultra-violet photons have more energy than infra-red photons.

When light photons fall on the semiconducting materials, such as the compounds of gallium and germanium, in opto-electronic receivers, electrons are released. Phototransistors and photodiodes, which are occasionally used, have high electrical resistance in the dark so that when they receive light they 'switch on' the electric supply. Light-emitting diodes operate by the reverse process. They receive electrical energy and convert it to light, often infra-red. Lamps receive electrical energy and, with varying amounts of efficiency, convert it into light energy — usually visible but occasionally ultra-violet with some infra-red.

By combining the various elements in the definition of light, a summarised definition can be formulated. Light is an intangible, fast moving, compound form of energy created by electrical energy

TABLE 1.2

Type of light	Frequency, pulse/sec	Source
Infra-red	9.0×10^{11}	Vibrating molecules and atoms
Visible	4.3×10^{14}	Vibrating electrons in atoms
Ultra-violet	7.3×10^{14}	Vibrating electrons in atoms

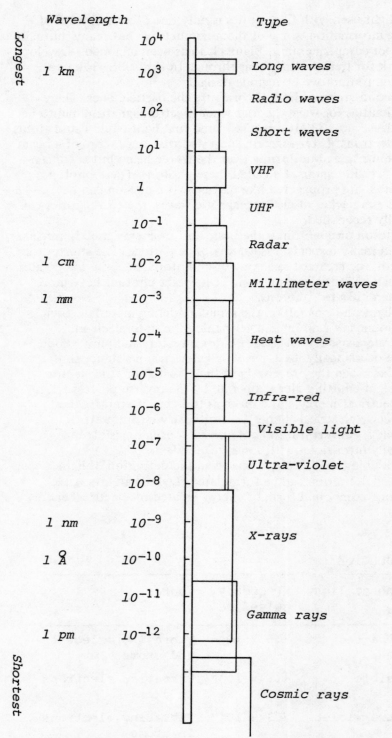

Figure 1.3 Energy forms and wavebands

and which can, with the aid of suitable devices, be returned to the
electrical form of energy and also initiate chemical reactions.
Initiation of a conversion process is one of the three functions of
light. The others are information conveyance and collection.

1.3 The functions of light

The functions of light are collection, conveyance and conversion.
If these three are combined they add up to one summary function —
communication. Communication is effected by other forms of
energy as well as light so the functions of light are described, in
this section, under the headings given in the first sentence.

Light *collects* information from an object or target. The light
may be emitted by an independent configuration of optical fibres
or the optical fibres integral with a fibre-optic instrument. The
return path to the receiver may be entirely through air, or through
several different media — air, glass or clear plastic arranged in an
optical system. Lenses, mirrors, prisms and optical fibres are components available for selective assembly into optical systems.

The object or target may be static or moving. Generally objects
viewed by independent lightguides and fibre-optic instruments are
static. Targets presented in fibre-optic sensing instruments are
frequently fast moving.

Information on the target is collected because light is composed
of a number of different wavelengths which give fine definitions to
the different colours of an object. Targets for sensing may collect
their information by being sensitive to contrasts. The most common
contrast is between black and white as in a bar code but there are
others. Data processing uses both holes and marks. Black marks
absorb more light than white ones.

The object is frequently in pictographic form — a burr inside a
precision component or a corrosion mark, when fibre-optic viewing
instruments are used. The light collects an image for conveyance to
the receiver.

A target for a sensing instrument is in analogue form. It does not
need to be in pictographic form because it is not intended to be
humanly perceived.

Both targets and objects vary considerably in size. Since one of
the properties of optical fibres is working to close limits the
variations are in degrees of 'smallness'.

Once the information has been collected it is *conveyed* between
two points. Light is obtained for direction to the object or target
from a light source. Reflection and refraction in fibre-optic
components assist this direction. From the object or target, the
light is conveyed by the passage of light waves to the receiver. The
receiver may be the human eye, or an instrument. The different
receivers respond at different speeds to the light they receive.

Electronic receivers respond more rapidly than the human eye but their response does not match the speed of light.

Fibre-optic telecommunications systems also use the conveying function of light. Light, often from a low-power laser, is presented to the optical fibres, in pulsed form. The pulse groupings represent the speech or electronic input presented to the encoder. The light is then conveyed through the optical fibres to the receiver for reconversion at the encoder.

Conversion is the third function of light. The conversion is into a more convenient form of energy. The first requirement of tele-communications is to have the light converted into an electrical form for decoding, into the audio or visual form — the end product.

Nerve cells in the human eye convert the light so that the image conveyed by the light, e.g. through an eyepiece, can be perceived by the brain. The human eye can distinguish more than 11,000,000 shades of colour. Light-sensitive materials in sensing instruments and systems convert the light into an electrical energy form.

The function of light is to initiate conversion in conjunction with an interface, be it the human eye, or a light-sensitive material or chemical. Intensity of light as well as the wavelength is a prerequisite for discharging this third function. Ultra-violet rays initiate a chemical reaction, as does visible light when used for photography. Figure 1.4 shows the energy, equipment and fibre-optic functional link-ups. The chart anticipates the chapter on fibre-optic instruments. It does not include specific applications because its purpose is to relate groups of circumstances in which fibre-optics can be used.

Fused optical fibres have been omitted in order to avoid impeding the use of the chart by adding a surfeit of detail. The electrical energy input is so extensively used with fibre-optics as to be almost universal. However, the fused-fibre otoscope can operate by collecting ambient light.

The headings for the four columns are spatial but they could reflect the three fibre-optic roles — illumination for viewing, non-visual and visual sensing and telecommunications. Columns 1 and 2 are the link-ups for discharging the role of illumination for viewing. Column 3 is telecommunications, while Column 4 is for non-visual and visual sensing.

In the first two columns the light has to pass through air as it does in the last column, even if the space between the probe and the target is only a few millimetres. Sensing instruments operate on the no-contact principle so there must be a small air gap. On the outward journey the light has also to pass through the glass quartz or clear plastic materials from which fibres can be made.

On the return journey, besides air, the light may have to pass through glass lenses or coherent optical fibres, depending on whether a rigid or a flexible instrument is used. If no fibre-optic instrument is used, the information returns to the human eye through air. The information for sensing is returned to the receiver through non-

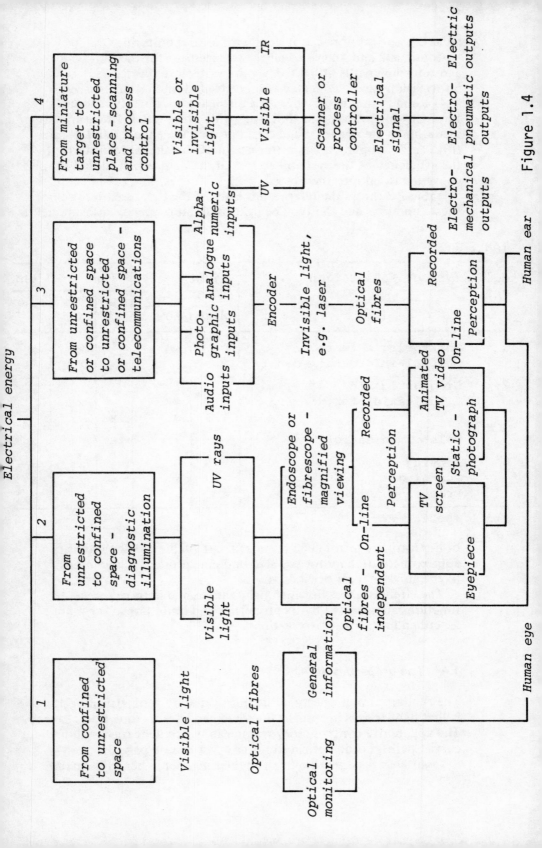

Figure 1.4

coherent optical fibres, as it does with local monitoring.

Columns 2 and 4 have a hairpin arrangement for the outgoing and returning paths of light. They are very close together so information can be collected from restricted or confined spaces.

Column 3 does not have a path through air. The only medium through which it passes is that from which the optical fibres are made, since the system is enclosed between the encoder and decoder. If it is not, coupling losses introduce inefficiencies which interrupt the efficiency of the system because all the light which should be conveyed is fed into the fibres.

Table 1.3 shows the relative intensity of the light used to discharge these functions and the type of light used. More precise measurements

TABLE 1.3

Col. ref.	Type of application	Light intensity		Type of light		
		High	Low	UV	Vis.	IR
1	Monitoring a function		x		x	x
	General information display	x			x	
2	Diagnostic aid –					
	illumination for viewing	x	x*		x	
			x	x		
3	Telecommunications		x			x
4	Sensing				x	x
	Scanning		x			x
	Process control	x			x	x

*With special TV

of intensity are introduced in subsequent chapters. High intensity approximates to a motorway sign or a ciné projector, low intensity to the glow on a car dashboard.

The circumstances in which light can discharge these three functions are affected by its properties and limitations. These are described in the next two sections.

1.4   The properties of light

The fundamental property, as demonstrated by geometrical optics, is that light travels in straight lines and can have its direction changed. The apparently contradictory manner in which light travels along curved paths, inside optical fibres, is described on page 31.

Light changes direction in an angular manner — hence the name

geometrical optics. The change of direction is governed by two laws
of light – reflection and refraction.

Reflection concerns opaque objects such as mirrors and the things
which are presented to fibre-optic instruments for viewing, or sensing.
Refraction concerns transparent lenses and prisms usually made from
glass but occasionally from clear plastic materials. These are incorporated in fibre-optic instruments and systems.

The word 'reflection' is derived from the Latin for bending back.
When light strikes a regular flat opaque specimen, it bounces off at
the same angle as that at which it met the reflecting surface. In
practice, however, not only a single ray but a bundle of rays are
simultaneously presented to a reflective surface as a cone or shaft of
light. The aperture angle of optical fibres may be between 15° and
130°. Some fibre-optic instruments have object lenses with an angle
as wide as 50°; some objectives, the other name for object lenses, are
even larger. The aperture angle is really a cone.

Unlike mirrors, objects which are viewed or sensed are neither
regular nor flat. Each small area of an irregular surface acts as a
reflector. The cumulative effect is that the object acts as a light
diffuser.

Reflection is used with parabolic mirrors to direct intense quartz
halogen light, inside fibre-optic light sources, onto the end of lightguides. Different objects reflect different amounts of light. Carbon
deposits on the crowns of pistons in diesel engines reflect as little as
4 per cent of the light they receive. The low level of reflectivity is a
function of both the colour and the surface texture. However, mirrors
absorb some light. The face is optically dense so about 4 per cent of a
ray's energy is expended in releasing itself from the face, i.e. 96 per
cent is reflected.

There are two different forms of reflection – internal and external.
The description so far applies to external reflection. *Internal
reflection* is the means by which light is transmitted through optical
fibres. The property of internal reflection is embodied in prismatic
binoculars, single-lens reflex cameras and range finders – and optical
fibres.

Total internal reflection, ignoring absorption by the reflecting
medium, takes place when the angle of incidence (striking angle)
exceeds 90° (see Figure 1.5). There are four rays of light leaving
a dense and entering a less dense medium, e.g. two kinds of
optical glass of different density, as with optical fibres. The first
ray of light, at right-angles to the interface, passes through the
less dense medium unchanged in direction; the same would be true
for light travelling in the opposite direction. Ray 2 shows the relation
between the angle of incidence and reflection. Ray 3 shows that for
a critical angle the ray is refracted parallel to the interface between
the media. For angles greater than the critical angle the ray is totally
internally reflected at the interface as shown by Ray 4.

Refraction is the other property of light; it is concerned with

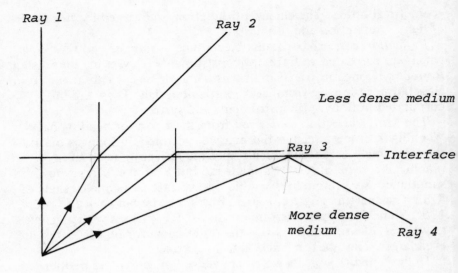

Figure 1.5 Total internal reflection

translucent substances or media — air, glass, water. The description
on Ray 3 of Figure 1.5 shows some similarities with total internal
reflection. Unlike those in solid substances, molecules in glass are
confined to a particular energy level. They cannot absorb and re-
emit light energy in the form of photons, by moving from one energy
level to another, and vice versa. Light energy consequently travels
through glass instead of being reflected as it would be with an opaque
substance. As the light from the less dense medium, e.g. air, strikes
the denser one, e.g. optical glass, its speed is retarded so that the
light changes direction. The angular and linear momentum of the
light are both affected. This property of light is reversible. The
effect of curvature on the directional change is described in
conjunction with lenses on page 66.

In the same way as reflective surfaces absorb some light, so some
refracted light is absorbed by ostensibly transparent media. There
are far greater and significant losses when light strikes a surface of
low reflectance. The thicker lenses needed for higher magnification
absorb more light than the thinner ones used for low-power
magnification.

Ray diagrams for lenses show how the light is refracted so as to
enlarge images, collect and collimate light. These are found in the
technical books on light physics and introduce such concepts as
real and virtual images.

The extent to which a translucent substance changes the
direction of a ray of light is expressed as a *refractive index,* e.g.
1.33 for water, 1.63 for the core and 1.52 for the cladding of
optical fibres made from optical glass. If different translucent

substances are incorporated in a system, then their refractive indices must be matched in order to minimise optical inefficiency.

Some filters are selective refractors. They absorb some wavelengths of light while allowing others to pass.

Besides the directional properties, *ultra-violet light* has important properties for fibre optics, particularly when used for biological applications. Between wavelengths of 118 and 153 nm little light is absorbed by organisms so it has little effect on them. Penetration is low; it only produces surface effects. Sunburn is an undesirable surface effect. Without introducing extraneous chemicals it can selectively arrest certain cell activities. The effect on the cells can be reversed by simultaneously or subsequently exposing the cells to the light in the near-UV region or in the longer ultra-violet wavelengths. In fluorescent dyes and inks, invisible ultra-violet light produces a visible reaction which is useful for certain types of crack detection. (Ultra-violet light, transmitted through quartz fibres is being used for hardening fillings in teeth.)

At the other end of the spectrum is invisible *infra-red light*. This light is hot and many opto-electronic devices respond efficiently to infra-red light in small quantities. This property is therefore one of the reasons accounting for the preference of IR-emitting and receiving electronic devices.

Infra-red light in the quantities produced by some of the larger lamps is a serious limitation on the ability of light to discharge the three functions enumerated above efficiently. There are some other limitations of light.

1.5   The limitations of light

The limitations of light are three — absorption, heat and diffraction. The limitations associated with fibre-optic components which must be additionally considered are described in Chapter 2 in the relevant sections. As with infra-red light, some of these limitations are, in some applications, principally opto-electronic sensing ones, used to advantage. In fact for those applications, instead of being limitations, they could be regarded as properties of light.

Absorption by air through which the light passes and inefficient reflection by an object are problems frequently encountered when illuminating for viewing. As the distance between the end of a light-guide or fibre-optic instrument and the object increases, so the intensity of light decreases. If the air is moist, the globules of moisture diffuse the light so that the quantity of light is reduced.

Although fibre-optic viewing instruments can view objects at almost telescopic distances through their lens systems, in ambient light, the absorption of light and the inaccessible spaces for which those instruments are made renders viewing without the assistance of artificial light impossible. Aperture angles of the fibres which conduct the artificial light are such that they illuminate an area greater than that of an instrument's object lens. The amount of light supplied by the light source to the system is not all efficiently directed. In some circumstances, collimation of light is required but this is not possible inside a fibre-optic viewing instrument, only with an indepedent lightguide.

Infra-red light, while it may be a convenient form of light for fibre-optic and other scanning systems, heating food and providing heat curtains in factories and multiple stores, constitutes a limitation of light for some other fibre-optic purposes. When light of high intensity is required, the intensity is not only in the visible part of the spectrum but also in the hot, infra-red part. The *heat* is in some cases sufficient to damage the cement between glass optical fibres and the parent material of ones made from synthetic materials. The damage usually results in low transmission of light so the fibres are unable to discharge the three functions of light — collection, conveyance and conversion.

Diffraction is not a serious limitation, but it may be encountered. It occurs when a beam of light encounters a sharp edge or slit. The beam is broken up into a series of light and dark bands. Bubbles in lenses would cause diffraction. This is one of the lens manufacturing problems which imposes a minimum diameter on the size of endoscopes. There are two kinds of diffraction — Fresnel and Fraunhofer.

The application described in Section 4.12 turns this limitation of diffraction to advantage in a quality application of non-visual sensing.

1.6 The quantification of light

Previously, relative as well as absolute terms have been used. Wavelengths of light and refractive indices have been quoted but no precise measurement of light intensity, reflectance and the other units in which light can be measured have been described.

Some of the concepts of photometry — the measurement of light — are abstract, so some analogies have been used.

Basic to the whole system of light measurement is the response of the human eye. This response is incorporated in the system of measurement of the *Commission d'Eclairage Internationale (CIE)*. The measures of light, relevant to fibre-optics, may be linked together in the following sequence; the succeeding element is proportional to the preceding one:

> Light energy or flux emitted
> Brightness or luminous intensity
> Light or illumination received
> Light reflected or illuminance

The measures for these elements are:

Flux (lumens)
Brightness (candelas)
Light received on an area (lux)
Light reflected (foot lambert)

Losses within optical fibres are also measured in decibels per kilo-metre or mile or expressed as a percentage of total input. The precise definitions are on page 36ff.

These measures are usually associated with fibre-optic light sources used for illumination and process control.

The *brightness* is dependent upon the wattage and type of lamp or light-emitting diode. The area a lamp can illuminate is an important consideration. Table 1.4 shows the connection between wattage, light output and area for three typical quartz halogen light sources. The lens system in the 500-W light source absorbs light from a larger lamp but the light is homogeneous.

Illuminance or *light reflected* introduces the other first-order variable. It is the measure of light leaving a surface, object or target.

TABLE 1.4

Energy, W	Light output, lx*	Area	
		mm²	in²
20	700,000	4	0.157
150	2,000,000	13	0.512
500	1,500,000	13	0.512

*Illumination is usually measured on a source at the face where a lightguide is attached. Table 1.5 quotes some data for familiar light sources.

TABLE 1.5

Type of source	Illumination, lx
Candle	1.20
100-W domestic incandescent lamp	1,200.00
1.5-m fluorescent tube	5,000.00
Sunlight	10,000.00
150-W quartz halogen lamp	2,000,000.00

TABLE 1.6

Place	Illumination, lx
School classroom	300
Factory, minimum	60
Factory – fine work, normal	450
Garage – normal	50
Stairs	50
Fabrication shop	80

TABLE 1.7

Material	Light reflected, %
Silvered surface	94
Polished aluminium	90
Fresh whitewash	90
Ceiling paint	80
Porcelain	75
Drawing paper	70
Bright chrome	70
White linen	65
Nickel	60
Matt steel – bright	55
Matt brass	50
Black paper	5
Black velvet	1

In order to reinforce the quantities with concepts, Table 1.6 has been prepared to show the light-received levels in some typical workplaces.

The inverse square law describes the relationship between the illumination of a surface and its distance from a point light source. The illumination received is inversely proportional to the square of the distance from the source. The law is modified if reflectors are used.

Table 1.7 is a table of illuminance values for comparison with objects which have to be illuminated for viewing.

When light sources with a continuous spectrum are specified, an additional quantity, besides area and intensity, appears — colour temperature, which is measured in kelvins (degrees Kelvin). It is a convenient way of describing the spectrum of radiation. The colour temperature of a non-black body is the temperature at which a black body, e.g. a piece of carbon, must be maintained in order that the distribution of energy in the spectrum of the non-black body may be matched as closely as possible. If the non-black body is radiating very nearly as a black body the colour temperature will be near the true temperature. If it is not the colour temperature will be of more practical value. The glow of metal changes colour as more heat is applied. Steel, as it is heated, changes from a dull red to a dazzling white colour.

Coloured photographic film is sensitive to colour temperature, 3200 - 3400 K being common. The kelvin is the unit of thermo-dynamic temperature: 0 K = 273°C or absolute zero.

The colour temperature varies with electrical input. Small incandescent lamps have colour temperatures between 2000 and 3000 K. A 60-W, 220-V incandescent lamp operated at 100 V has a colour temperature of 2300 K. The colour temperature of the sun is 6000 K when the sky is blue; that of a candle and dawn sunlight is approximately 2000 K. A warm white fluorescent lamp is 3000 K and that of a 60-W domestic incandescent lamp 2800 K when operated at its correct voltage.

The *resolution of images* obtained by coherent optical fibres, or electronic means, is quantified in terms of line pairs/mm. Figure 1.6 shows a test card which could be used with a fibrescope or television system. The resolution of optical glass lenses is continuous and not made up of a series of dots.

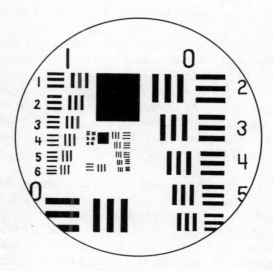

Figure 1.6 Test card transmitted via a fibre-optic system

TABLE 1.8

Units	1 lm/m^2	1 lm/cm^2	1 lm/ft^2	Term
1 lm/m^2	1.00	0.0001	0.093	lux
1 lm/cm^2	10,000	1.0	930	phot
1 lm/ft^2	10.76	0.001076	1.00	foot-candle

TABLE 1.9

Units	cd/cm^2	cd/ft^2	lm/cm^2	lm/ft^2	Term
cd/cm^2	1.0	930	3.142	2,919	stilb
cd/ft^2	0.0011	1.0	3,400	3.14	
lm/cm^2	0.318	342	1.0	929	lambert
lm/ft^2	0.000342	0.292	0.0011	1.0	foot-lambert

The instrument is focussed on the test card. The point at which a clear image is obtained is read off from the chart; 25 line pairs/mm would be a poor resolution while 40 line pairs/mm would be good.

Since there are a number of different quantities in use, Tables 1.8 and 1.9 convert illumination and luminance units, respectively. (The abbreviation for lumen is 'lm' and that for candela, 'cd'; the rest should be familiar.)

The precise definitions of the terms used are as follows:

Lumen (lm)   The lumen is the luminous flux emitted within unit solid angle of 1 steradian by a point source having uniform intensity of 1 candela.

Steradian (sr)   The steradian is the solid angle which, having its vertex in the centre of a sphere, cuts off an area on the surface of the sphere equal to that of a square having sides of length equal to the radius of the sphere.

Candela (cd)   The candela is the luminous intensity, in the perpendicular direction, of a surface of area 1/600,000 m^2 of a black body at the temperature of freezing platinum under a pressure of 101,325 newtons per square metre. A light source has a luminous intensity of 1 cd if it emits, in 1 sec, 1/60th of the light energy radiated in the same period of time from a long cylinder of thorium oxide kept in a bath of solidifying platinum and observed along the direction of the axis.

Lux (lx)   The lux is equal to an illuminance of 1 lumen per square metre.

The depth of focus is also used to quantify light. It is the distance through which an object or a viewing instrument can be moved without the image becoming blurred.

The next chapter describes the components used for discharging the functions of light. Electrical devices create it, optical devices collect and convey it and electronic devices convert and interpret it.

Two

BASIC FIBRE-OPTIC COMPONENTS

2.1 Résumé

The eight groups of components used in fibre-optics are described in terms of the associated fibre-optic instruments and equipment, their physical forms, properties and limitations. The latter are allocated appropriate sub-sections. Where components, and this is the majority of them, have applications in other technologies, the description has been directed towards their fibre-optical uses. (Lenses for telescopes and filters for photographic cameras have not been described.) Components in the eight groups which have no fibre-optic applications have been omitted. Street lighting is not described in Section 2.7 nor hi-fi in Section 2.8. Optical fibres are used in all the fibre-optic instruments described in Chapter 3 except light sources. They also have independent uses in various configurations. In view of their central importance, additional sub-sections have been included in the section devoted to optical fibres.

Many optical fibres are made from the same materials as lenses so a description of optical glass appears in Appendix 2. The chemistry of polymer materials is somewhat complex so there is no similar appendix for them. Electronic devices have their own groups of semiconducting raw materials — gallium, germanium and silicon compounds being the principal ones but there are others. Mirrors sometimes use metals and chemicals of several types appear constantly in connection with the different components. In the manufacturing hierarchy, lenses, prisms, mirrors and filters are semi-fabricated components. Electronic devices, lamps and lasers are minor assemblies. Optical fibres are like their optical counterparts — a semi-fabricated component.

2.2.1   Filament conductors

Figure 2.1 shows a straight ray of light entering a curved optical
fibre magnified approximately 400 times. The principle by which
the light travels inside the fibre is that of total internal reflection.
The ray enters the fibre at a shallow angle. Depending on the type of
fibre the angle varies between 15 and 130°. Until it strikes an interface between the fibre core and its cladding, the ray is not redirected.

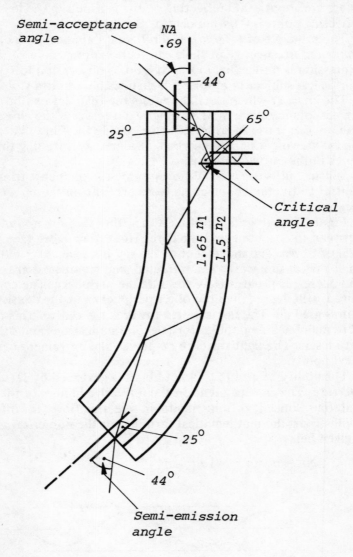

Figure 2.1 Ray of light in clad optical fibre

It travels inside the core at a small angle relative to the optical axis. At the interface the ray adjusts itself to the curvature with the fibre core and is reflected. There are limitations on the bending radius of fibres (see Section 2.2.4).

At the interface, total internal reflection takes place since the refractive index of the cladding is greater than that of air but less than that of the denser core glass. Radiant energy is trapped inside a totally reflecting guideway provided the refractive index exceeds 1.4.

Total internal reflection continues beyond the end of the fibre. A fibre 1000 mm long has reflectance to a length of about 1060 mm. There are about 1000 reflecting points per metre. The loss per reflecting point is 0.04 per cent of the light received.

The principle of *total internal reflection* applies to *all optical-fibre filaments*, irrespective of their end section dimensions or shape. The expression is misleading in that it ignores absorption losses. Infra-red light is less subject to absorption than visible or ultra-violet light.

The angle at which the light enters the filament or fibre is called the acceptance angle; that at which light leaves is the emission angle. The ray can travel in either direction inside the fibre. When illuminating for viewing, the light can be considered as travelling through fibres in the outward direction.

Sensing units use coaxial fibres; some to conduct it from the emitter to the target and some to return it from the target to the receiver.

Figure 2.2 shows a ray of light in a fibre that has *no interface* between the cladding and the core. These are a more recent generation of fibres. The movement of the ray is less angular and resembles an electrical sine wave. The refracted angle is greatest at the centre and decreases progressively towards the periphery. The core of the fibre is still denser than the outer periphery but the density transition is gradual. The ray is pulled towards the centre. This type of fibre may become popular for telecommunications and other uses with lasers. The light travels by a pincer-like movement between focal points.

The ability of a fibre to accept light is governed by its *numerical aperture.* This is dependent on the refractive indices of the fibre materials. Since it is fundamental in selecting fibres for different applications the mathematical formula for the numerical aperture is given below:

$$\text{NA} = n_0 \sin x_0 = (n_1^2 - n_2^2)^{\frac{1}{2}}$$

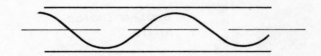

Figure 2.2 Ray of light in cladless fibre

where NA is the numerical aperture, n_0 the refractive index of ambient air, n_1 the refractive index of the fibre core, n_2 the refractive index of the cladding and x the angle at which the ray can strike the polished face of the fibre.

Assuming an angle of incidence of 36° for the ray and substituting typical values in the formula:

$$NA = (1.63^2 - 1.52^2)^{1/2}$$
$$= (2.659 - 2.3104)^{1/2}$$
$$= 0.3486^{1/2} = 0.589$$

The description of the conduction of light through the filaments or fibres has partly anticipated the following sub-section on structure to some extent.

2.2.2 The structure of optical fibres

There are two types of optical-fibre structure — cladded and cladless. The former are extensively used for illumination in the viewing and sensing roles of fibre optics. In both types of fibre the denser transparent material constitutes the core.

A ray of light reflected at a cladding/core interface penetrates into the cladding for a distance equivalent in length to a half a wavelength of the light that the fibre is conveying or transmitting. The cladding cannot therefore be less than half a wavelength in thickness. A restriction on the thickness of the cladding is imposed by the need to pack as many fibres as possible into the available area. A 0.005-mm (1.9 μ) diameter fibre would have a cladding thickness of 0.0003 mm (1 μ). As well as reflecting light, the cladding affords some protection to the core. In certain circumstances, principally with thick mono-fibres made from synthetic materials, the cladding is peeled off near the end of a lightguide. This bleeds light from the periphery of the fibre which augments that emitted from the end.

Experiments have been made with optical fibres for monomode

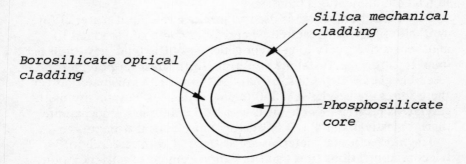

Figure 2.3 Phosphosilicate fibre

telecommunications in which the core is thinner than the cladding, e.g. 0.013 mm (50 μ) for the cladding but only 0.0003 mm (1 μ) for the core. In the region 650 - 780 nm this fibre is more efficient than fibres with thin cladding but the coupling problems are greater. The core has room for only one wave of light and can alter waveforms.

Multimode telecommunications are using, amongst others, cladless fibres. However, phosphosilicate fibres have two claddings — an optical and a mechanical one (see Figure 2.3). They have very low attenuation — 2 dB/km (3.2 dB/mile).

2.2.2 Optical-fibre materials

There are two groups of materials from which optical fibres are made — natural and synthetic. The distinction is conventional but imprecise because the synthetic materials are derived by polymerisation from natural ones. However, the extensive use of the word in connection with textiles, to differentiate the more complex polymer and acrylic yarns from natural wool, has eliminated the possiblity of confusion.

The natural raw materials are often silica-based. Silica (SiO_2) has many forms and compounds. In its eroded and deposited form of sand, it is used as a raw material for the manufacture of some optical glass (see Appendix 2). Quartz (also SiO_2), in the transparent milky crystalline form, is used when ultra-violet light has to be transmitted because glass will not transmit such short wavelengths. Synthetic fibres are more efficient than glass in this respect. Quartz fibres are brittle because they break along the edges of the crystals. They have a loose Teflon cladding because there is no suitable glass for bonding.

Synthetic fibres are either extruded or teased from polymer materials; there are different brand names. One type has a polystrene core clad with bonded transparent methyl metacrylate which can, like yarns, be dyed. Another type has a polymethacrylate core (hence the generic term acrylic) with another polymer cladding of lower refractive index. Polystyrene is also used as the strength member in telecommunications bundles.

Polymerisation, which is used to produce the input material for synthetic fibres, is a reaction whereby one molecule is added to another to give a very long chain which has little tendency, unlike quartz, to form a crystalline structure. In this respect polymers resemble glass which is a super-cooled liquid. The raw materials for these fibres are acrylic acid and methacrylic acid. Polymethylmethacrylate is regarded by chemists as a hard colourless, transparent, thermoplastic material of considerable mechanical strength.

The more unusual materials are associated with research into telecommunications fibres and the improvement of anti-radiation resistance of glass fibres for transmitting visible light (page 42). Before the development of phosphosilicate fibres, a fine capillary

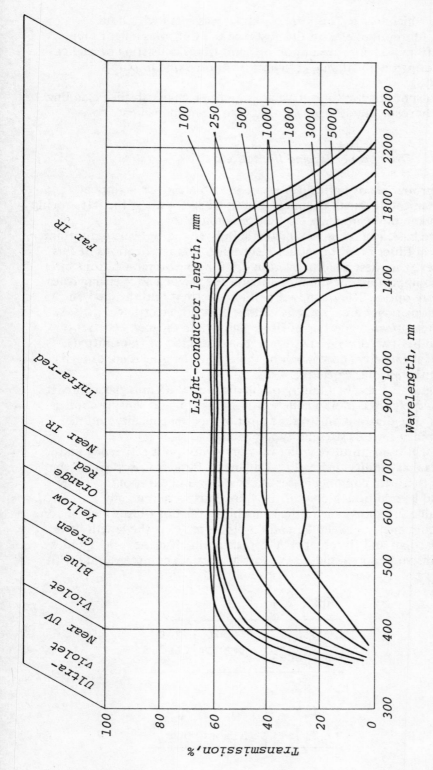

Figure 2.4 Spectral capacity and capability of popular glass fibres (initial losses included)

tube, which was to form the cladding, was filled with liquid
hexachlorobutadiene: in this instance the fibre was a light pipe!
This fibre was improved upon by solid fibres consisting of a silica
core doped with titania, germania or boric oxide in pure silica
cladding.

Phosphosilicate fibres have an impure mechanical silica cladding
then borosilicate optical cladding with a phosphosilicate core.

2.2.4 Their properties and limitations

The properties of optical fibres are in six categories — optical,
mechanical, thermal, chemical, electrical and sonic — the latter being
the paramount property for telecommunications.

The basic *electrical* property is simply stated but very important.
Optical fibres do not conduct electricity. The implications of this
property are described in Section 4.5 for telecommunications but it
is also important for the illumination for viewing and sensing roles
of fibre optics. They conduct neither electricity nor heat so, in the
object or target area, there is neither heat not electricity.

The *optical* properties of fibres have three aspects: spectral
capability (what wavelengths can be conducted or transmitted),
spectral capacity (how much of those wave-lengths is conducted),
and their angular capacity.

Figure 2.4 shows a family of curves for a 0.05-mm glass fibre. It is
extensively used in non-coherent illumination and sensing applications. The diagram illustrates both the spectral capacity and its
capability for this size and type of fibre.

It will transmit ultra-violet rays only with poor efficiency and its
spectral capability decreases with length. Besides losses within the
fibre there are coupling losses at the entry and exit points. There
would be additional losses if the fibre ends were optically untreated.

Table 2.1 shows the optical capability for this fibre which has an
aperture angle between 55 and 65°. This restricts the length of
endoscopes and fibrescopes. For maximum efficiency, lightguides
and instruments should be as short as possible commensurate with

TABLE 2.1

Length		Input light transmitted,%
m	inches	
1	39	50
3	117	35
8	312	Almost 0

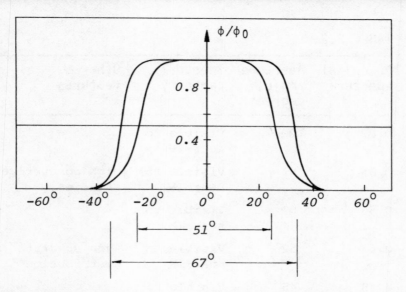

Figure 2.5 Polar or re-radiation diagram for popular glass fibres

the application. There are constant initial losses of approximately 38 per cent. These losses are attributable to Fresnel reflections and interstices between fibres.

Figure 2.5 is a polar or re-radiation diagram. The numerical aperture could be 0.589 and the acceptance angle 72°.

The scope of fibre optics is large so applications are encountered where there are requirements to transmit wavelengths of light unsuitable for this fibre. The use of quartz and synthetic materials for ultra-violet rays has already been mentioned. Different aperture angles are required. Telecommunications have fewer launching or coupling problems if the angle is small but illumination and sensing applications prefer a larger aperture.

Table 2.2 shows the connection between numerical aperture and the aperture angle. All the fibres transmit visible and some infra-red light. They are not all available in all the different diameters. Each one of the above fibres could have a spectral capability graph similar to Figure 2.4 and a polar diagram (Figure 2.5) giving more precise information than columns 3 and 4. Figure 2.6 is a spectral capacity and capability diagram for quartz and Figure 2.7 for a synthetic fibre. In neither case can the aperture be specified because the dimensions cannot be held as tightly as they can with glass. Synthetic fibres with a bonded cladding as opposed to a varnish one have the numerical aperture specified. Figure 2.8 shows the effect of coupling losses on a fibre extruded from a synthetic polymer material.

There are several differences in the mechanical and thermal properties which prevent fibres made from the different materials being interchangeable.

TABLE 2.2

Numerical aperture	Aperture angle, degrees	Spectral capacity	Other features
0.87	120	Visible to near-IR	
0.65	81	Visible to near-IR	High conductive efficiency
0.56	69	Visible to near-IR	
0.43	52	Visible to near-IR	Even spectral distribution
0.38	45	Visible to medium IR	
0.21	24	Visible to near-IR	

The nominal *tensile strength* of a 0.01-mm glass fibre is 100 kg/mm^2. Tensile strength increases as the fibre diameter decreases. However, the jacket prevents the tensile strength of fibres being tested in practice. Some telecommunications fibre bundles have a strengthening member incorporated alongside the bundles inside the cable.

There are limits to the *bending radius,* but the jacket usually prevents this being exceeded. As thickness increases the bending radius decreases. For a 0.01-mm fibre, the nominal bending radius is 0.07 mm. These fibres are too fine to be used as mono-fibres. Table 2.3 gives the bending radii, in both metric and imperial units, of lightguides in a PVC jacket. PVC is the most flexible of jacket materials.

Fibres extruded from synthetic materials can be permanently bent by gentle heat. Fused glass fibres are bent and shaped by heating to 700°C.

Excessive flexing of fibres ruptures the cladding and reduces overall transmission efficiency by leakage. More serious, however, are fractures due to bending. The sharp edges at the fractured point contaminate adjacent fibres even to the extent of shearing them.

Some fibre fractures may occur during the assembly optical fibres. Upper limits for fracture percentages are often specified, e.g. 6 per cent for illumination and 3 per cent for coherent image-transmission systems.

Fractures, in addition to bending, can be caused by *flexing* although

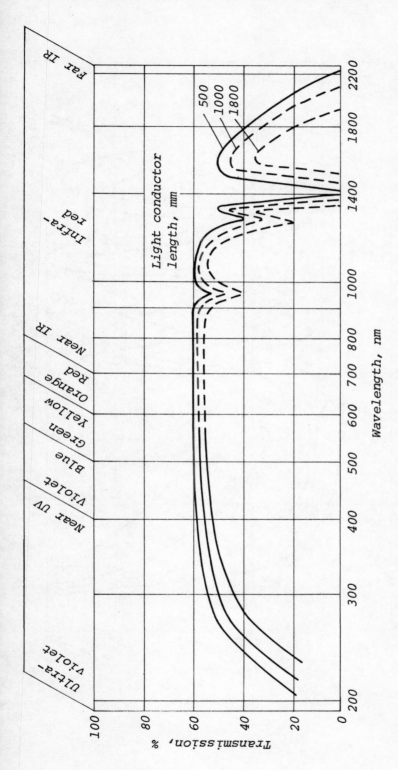

Figure 2.6 Spectral capacity and capability of quartz fibres (initial losses included)

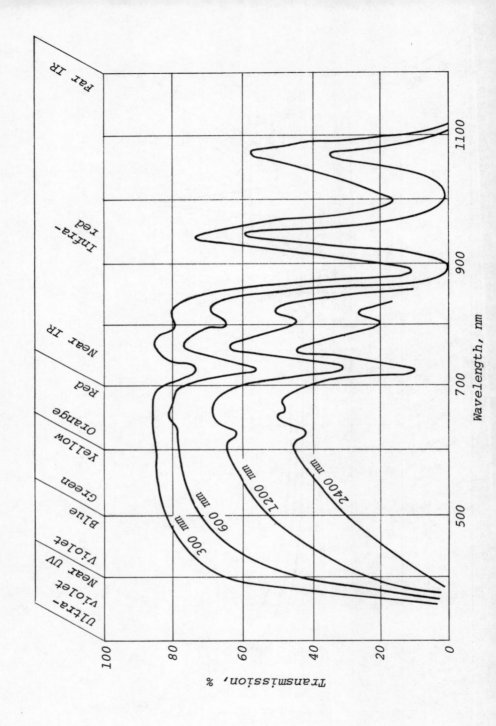

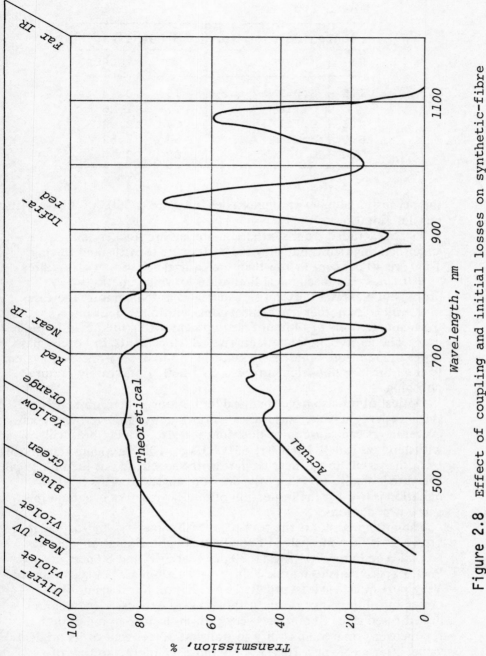

Figure 2.8 Effect of coupling and initial losses on synthetic-fibre lightguide 300 mm long

TABLE 2.3

Lightguide diameter		Bending radius	
mm	inches	mm	inches
1.5	0.059	19	0.74
3.0	0.118	32	1.26
4.5	0.177	64	2.52
6.0	0.236	64	2.52
9.0	0.354	89	3.50

the permitted number for lightguides is high, e.g. 500,000 flexes without deterioration.

Fibres extruded from synthetic materials are thicker and have a larger bending radius but they are better able to withstand flexing. They are 40 per cent lighter than the equivalent mass of glass fibres.

Brittleness is a mechanical limitation to which quartz is very prone, but glass is not exempt. It increases as temperature increases.

Hardness is another mechanical consideration which concerns the manufacturers of fibre-optic equipment, often unknown to their users. The multi-component adhesive which cements, or bonds, glass fibres used in fibre-optic instruments has to have a hardness equivalent to the fibres for polishing purposes. Synthetic fibres rarely require polishing.

Optical fibres also have chemical limitations. When subjected to certain energy wavelengths they are attacked by *radio-activity*. High-power lasers can cause some deterioration also. Some fibres will withstand up to $1 \times 10^7 r$ $(1 r = 10^{-2}$ J/kg $= 100$ erg/g) and transmit light from the visible to the near-IR part of the spectrum but the numerical aperture is small — 0.45 and the aperture angle 53°. This level of radiation is low for the inspection of nuclear reactors in submarines and power stations.

Radiation discolours the transparent fibre material by turning it brown. It is then incapable of conducting light. Radiation glass has less silica and more lead oxide; 20 per cent SiO_2 and 80 per cent PbO. Experiments using fused silicas and cerium are being conducted. Very pure quartz is least affected by radiation but it cannot be drawn into fibres very easily. Although very expensive, germanium-doped fused silica fibre shows very good resistance to radiation but the recovery time to an ability to transmit 50 per cent of its original value, after receiving a 1000 r X-ray pulse, is inferior to that of some fibres made from synthetic materials.

Besides radiation, *water* also affects fibres chemically and optically. An aperture angle of 120° is reduced by a third, to 80°, in water.

The *cement* between glass fibres is attacked if immersed for long periods. For autoclaving, a special multi-component adhesive is used.

Synthetic fibres are also subject to chemical attack by *water and solvents* such as ketones and esters as well as aromatic and halogenated carbons.

The *thermal limitations* of optical fibres are different for glass and fibres made from synthetic materials. Synthetic materials are rated from -40 to 80°C with an intermittent increase to 95°C. The heat need not necessarily be from the object or target; it could be infra-red light from the light source. Heat-absorbing filters have to be used to protect them.

Glass fibres are not damaged by heat until long after the cement is. The cement is rated for temperatures from -40 to 120°C. They also need protection from infra-red light. Fused fibres withstand 500°C.

Fibres can be used at high temperatures if a special outer jacket is fitted to the instrument. One fibrescope with such a jacket for air or water is rated for temperatures up to 1200°C. The jacket has to have a sapphire window in a titanium frame.

Telecommunications fibres have their properties and limitations expressed in terms of *sonic* efficiency, although light has the function of conveying the input from the encoder to the decoder. Since the use of optical fibres is envisaged for long-distance land lines, a loss

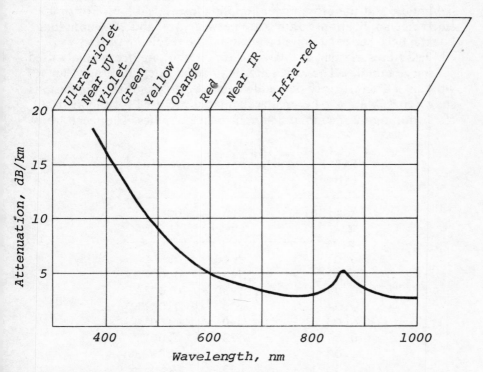

Figure 2.9 Spectral attenuation of phosphosilicate fibre (1.2 km long)

of 2 dB/km, although small, can be cumulatively unacceptable. Figure 2.9 shows the spectral attenuation for a 1.2-km (¾-mile) length of silica-clad fibre with a phosphosilicate core. If the end of the bundle is well polished, then launching losses should be minimised. The mechanical difficulties of end-shaping single-made fibres is impeding their adoption for telecommunications because the maximum amount of light energy must be launched into the system. The illumination for viewing and sensing roles of fibre optics are more tolerant of these losses. Telecommunications fibres are secondly specified in terms of time. A fibre with a 0.05 mm (50-100 μ) core diameter had a laser pulse speed of 3 nsec in 50 m (164 ft).

2.2.5 Their dimensions and shapes

There are three considerations — length, diameter and end-section.

Telecommunications fibres may be up to 1.25 km (0.78 mile) long. A standard endoscope is 2 m long with the additional length of 1.5 m for the lightguide which connects it to the light source. The overall length of a fibrescope, which has an integral lightguide, and therefore fewer coupling losses, is 4.5 m. For opto-electronic applications, probes are between 100 and 2000 mm long but the light has to 'double back' along the receiving fibres.

Glass fibres as small as 0.01 mm are rare but are occasionally used for image bundles. Their diameter is so small, they are almost un-workable. The majority of cladded glass fibres are of the dimensions shown in Table 2.4. The smaller diameters are used for the high resolution requirements of coherent image bundles. There can be as

TABLE 2.4   Diameters of glass fibres

Nominal mm	μ	Nominal inches
0.01	8–10	0.0004
0.02	15–20	0.0008
0.03	15–30	0.0011
0.04	45	0.0015
0.05	50	0.0019
0.07	75	0.0027
1.00	100	0.0393
1.50	150	0.0589
2.50	250	0.0982

many as 10,000/mm^2.

Above 1 mm, glass fibres are brittle and resemble glass rods; below that diameter, unlike glass, they are flexible.

Table 2.5 gives the diameters of a range of synthetic fibres. The increments indicate whether the fibre is made in a country using the metric system of measurement or not.

TABLE 2.5   Diameters of synthetic fibres

mm	0.127	0.254	0.508	0.762	1.016
inches	0.005	0.010	0.020	0.040	0.060

Phosphosilicate fibres for telecommunications have overall diameters between 0.05 and 0.10 mm. Self-focusing cladless fibres have diameters between 0.1 and 1.0 mm.

The majority of optical fibres are *cylindrical* in *section* but in order to improve compacting square section fibres have been drawn, as well as oval shapes. The interstices between fibres prevent all the space in a ferrule being occupied by fibres. In image guides, fibres occupy approximately 74 per cent of the available space. The image resolution for coherent fibres is inferior to a lens system which occupies all the available space.

The sub-section on assembled fibre configurations describes some of the infinitely large number of shapes into which these fibres can be collected and assembled.

2.2.6   Methods of manufacturing optical fibres

There are five manufacturing processes. All of them are complicated and all involve the use of heat and a spooling element.

1   The most widely used method for the manufacture of cladded glass optical fibres is that of drawing through a two-chamber nozzle (see Figure 2.10). The core glass is heated in the centre and the cladding in the outer chamber, to the state where it can be drawn through the nozzle. Heat is maintained and stress relieving effected by passing the drawn fibre through a heated ring before the spool where it is sufficiently cooled for each fibre to be quite separate. Drawing machines have banks of nozzles so up to 200 fibres are simultaneously spooled.

2   More precise dimensions can be held by having the core as a rod and the cladding as a tube. Before these two components are fused, air has to be evacuated because air bubbles would introduce diffrac-

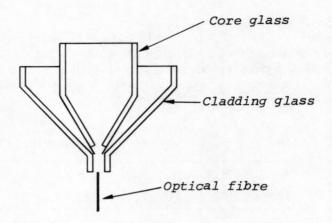

Figure 2.10   Two-chamber nozzle

tion into the end product. The assembled components are heated until they begin to drip. Then winding takes place similar to that in the first process. The winding speed must be constant because it affects the diameter of the fibres. Increasing the speed reduces the diameter, just as reducing the speed increases it.

3   Filaments of synthetic materials are suitable for extrusion through multi-holed dies which have an outer chamber for applying the sheath. The sheath may be applied as a varnish but is more likely to be another fused polymer.

4   Phosphosilicate bi-clad fibres, for telecommunications, are made by passing vapours of phosphorous oxychloride ($POCl_3$) and silicon tetrachloride ($SiCl_4$) and oxygen down a silica tube which need not be pure. This eventually forms the mechanical cladding. They are heated to 1500°C. Layers of phosphosilicate can then be deposited on the inside of the tube (see Figure 2.11). The initial layers, which become the optical cladding, can be made to have a lower refractive index than the core by using a small concentration of phosphorous oxychloride or boron trichloride. After deposition the tube is collapsed to a solid form for drawing into fibres.

5   Self-focusing fibres are made by a complicated process not unlike the puddling method of producing wrought iron, but in a heated nozzle. They are drawn in the same manner as the majority of glass fibres.

2.2.7   Assembled configurations

Although the title may suggest otherwise, fibres are not always

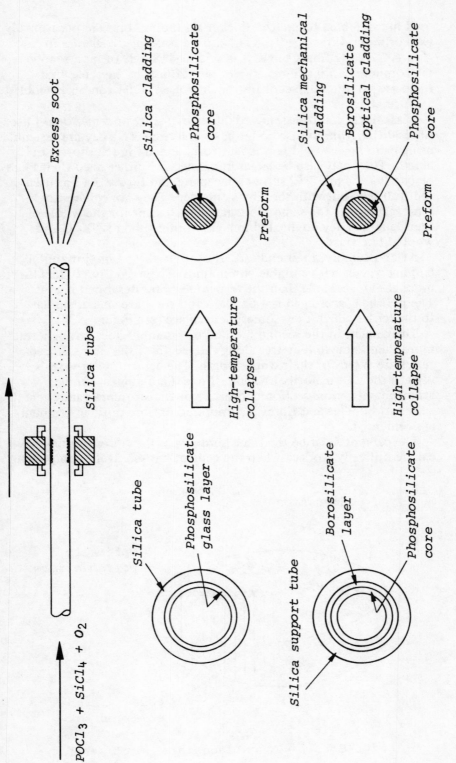

Figure 2.11 Phosphosilicate fibre

used in assembled form. The thicker synthetic fibres are occasionally used as monofibres but they can also be assembled. The use of fine glass and quartz fibres singly is rare indeed. Single or mono-mode telecommunications fibres are not monofibres in the sense that there are no other adjacent fibres. Only glass fibres can be assembled by fusing.

Glass fibres can be assembled either by *fusing* or *cementing*. Fused fibres differ from unfused ones in several respects. They are assembled using heat, have a rigid permanent form and are much shorter in length. They may also be larger in diameter. Neither need their faces be parallel. Figure 2.12 shows a faceplate and Figure 2.13 a fused rod, full size, but with the fibres enlarged. They are coherently arranged so that the same fibres remain adjacent for the whole of their length. They are heated in a glass outer tube to 700°C and worked like glass.

After assembly, their ends are optically polished on horizontal lapping wheels with suitable compounds. Their rigidity renders them more easily workable than the cemented fibres described below. The polished faces need not be plane but can have shapes similar to the lenses which they sometimes replace (see Figure 2.34).

The cement in the second method of assembly does not penetrate more than a few millimetres along the length of the fibres. Fused fibres are fused for their whole length. The cementing process is used in the vast majority of applications. Lightguides, annular arrangements, cross-section converters or transformers, random and coherent bundles and fibres incorporated in fibre-optic instruments, are cemented.

The cement must be the same hardness as the fibres so that they can be optically worked when the cement has set. Multi-component

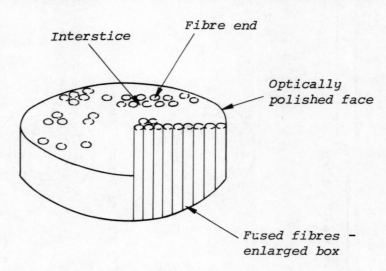

Figure 2.12 Fused faceplate - cut away

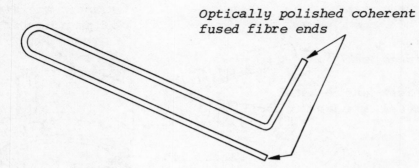

Optically polished coherent fused fibre ends

Figure 2.13 Fused rod - half size

adhesives based on epoxy resin are used for cementing glass fibres and acrylic varnishes for synthetic ones.

Synthetic fibres are, for some applications, used *uncemented* and loosely held in the jacket. When cement is used, a metal ferrule or plastic moulding has to be used in conjunction with the jacket. The fibres are drawn through the ferrule and the cement is distributed between the fibres. After curing, the fibres are trimmed with an abrasive wheel. They are then polished. The process of distributing the cement is aided by the addition of a visible dye. This dye frequently gives a coloured appearance to fibre ends.

Metal can also be used for housing the other shapes where the application environment inhibits the use of the more easily worked plastic mouldings for cross-section converters. The cement, besides holding the fibres rigid for polishing, to some extent restricts flexing at the ends, where fibres are most subject to damage. Polishing can be undertaken on a DIY basis for the less demanding applications.

Table 2.6 lists the more normal *end-section* shapes into which optical fibres are assembled. To these could be added all the other geometrical shapes — triangles, ovals, rhomboids, polygons and

TABLE 2.6

Form	Shapes		
	Circle	Square	Rectangle
Solid	x	x	x
Hollow	x (annulus)	x	x
Number of branches			
One	x	x	x
Several	x	x	x

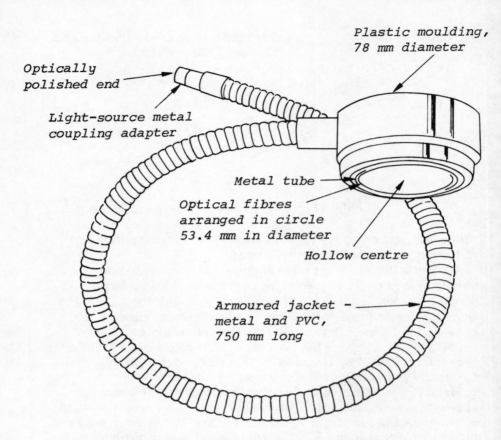

Plastic moulding, 78 mm diameter

Optically polished end

Light-source metal coupling adapter

Metal tube

Optical fibres arranged in circle 53.4 mm in diameter

Hollow centre

Armoured jacket – metal and PVC, 750 mm long

Figure 2.14 Ring light – complete circle

many more. The number of alternative assembled configurations is infinite. Solid circles are used for single lightguides for illumination. They can be used independently or attached to a fibre-optic viewing instrument to connect with its integral fibres which have a hollow arrangement. Probes for sensing have a similar arrangement. Figure 2.14 shows a ring light. This comprises a solid 13-mm diameter fibre bundle which is distributed in a 50-mm diameter circle inside the moulding with the fibres slanting inwards. Figure 2.15 shows a ring light with the fibres distributed round a pitched circle diameter.

Figure 2.16 shows a multi-branch lightguide and Figure 2.17 a multi-branch, multi-slit, cross-section converter. These are also known as shape converters or transformers. The solid circle is converted into other shapes.

Figure 2.18 shows a solid, slim, rectangular cross-section converter with a continuous 0.5mm slit, Figure 2.19 an interrupted-slit cross-section converter and Figure 2.20 a similar cross-section converter but the moulding retains the fibres permanently at 90°.

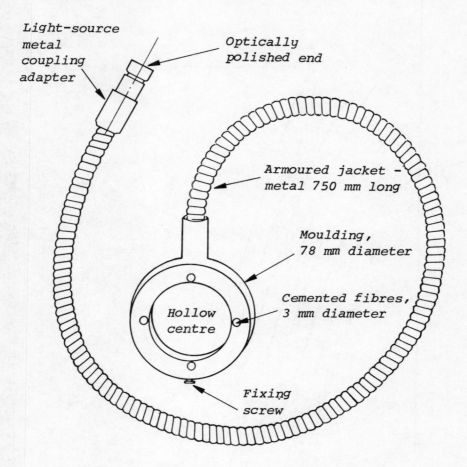

Figure 2.15   Four-point ring light

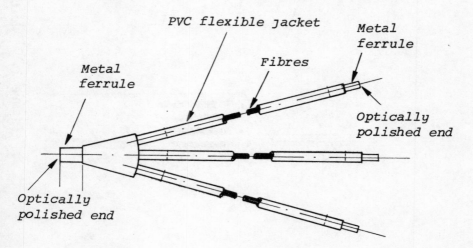

Figure 2.16   Multi-branch lightguides

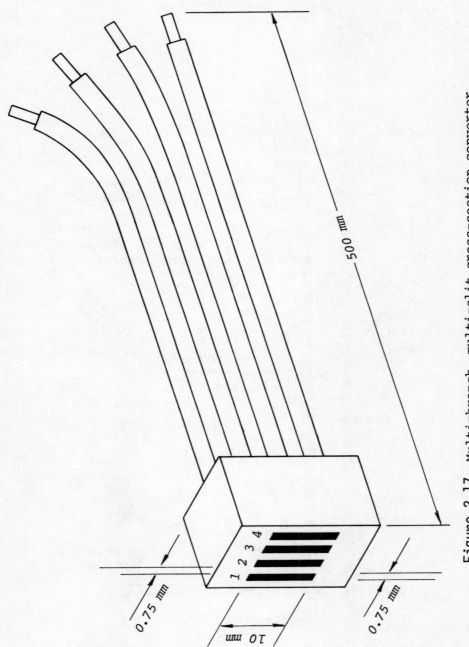

Figure 2.17 Multi-branch, multi-slit cross-section converter

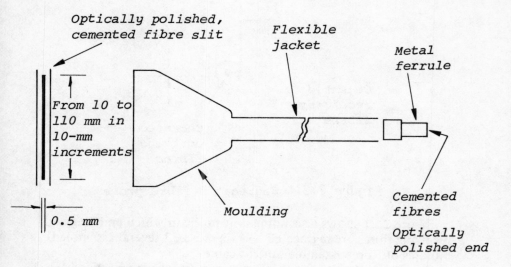

Figure 2.18 Cross-section converter with continuous slit

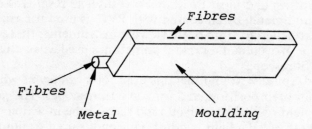

Figure 2.19 Interrupted-slit cross-section converter

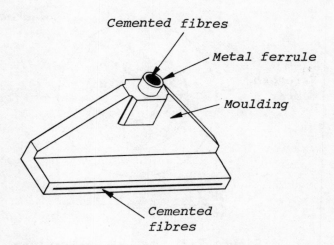

Figure 2.20 Right-angled cross-section converter

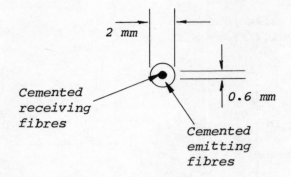

2 mm

Cemented
receiving
fibres

0.6 mm

Cemented
emitting
fibres

Figure 2.21 Segregated fibre probe

Figure 2.21 shows a scanning unit probe in which emitting and receiving fibres are segregated and Figure 2.22 several assembled bundles of fibres assembled into a cable.

Except for the thicker synthetic ones, fibres are assembled into a circular, hollow, protective, *jacket*. This counteracts some of the limitations of fibres. These jackets are made from plastic or metal. PVC is the softest and most flexible. Polyethylene resin is also used. Spirally wound metal, often coated with PVC, is used for armouring lightguides attached to fibre-optic viewing instruments; these have to be flexible. Woven metal braid is sometimes used as an alternative to PVC for fibrescopes.

Semi-rigid, spirally wound metal is used for swan necks which have to retain their position, and for some fibrescopes. The jacket can be a straight or slightly curved rigid metal tube in which case the assembly is called a light conduit. One end is used to illuminate small holes and the other end is a push-fit into a lightguide. Figure 2.23 shows a light conduit with a removable mirror.

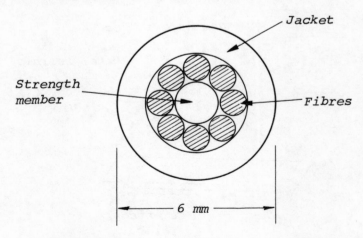

Jacket

Strength
member

Fibres

6 mm

Figure 2.22 Telecommunications
cable - enlarged

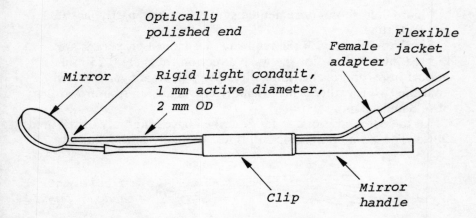

Figure 2.23 Light conduit with dental mirror

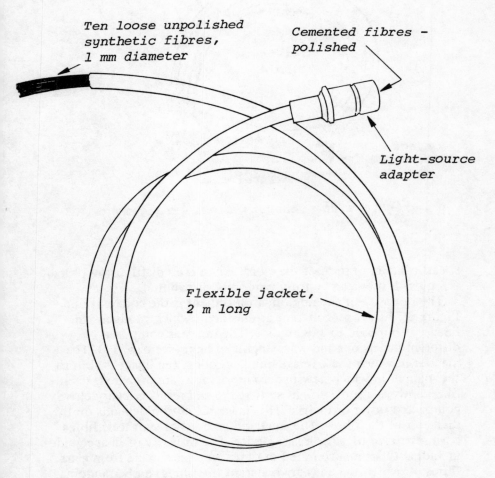

Figure 2.24 Synthetic fibres in PVC jacket

Figure 2.24 shows uncemented synthetic fibres partly jacketed in a PVC tube.

Synthetic fibres with the requisite mechanical properties are knitted into ribbon. For the warp direction there are 0.25-mm optical fibres and in the filling direction polyamide yarns which hold the two together. This assembly is available in ribbon widths of 6, 8 and 12 mm.

All the configurations so far described have randomly distributed fibres. The *'random'* description is not normally in the limited

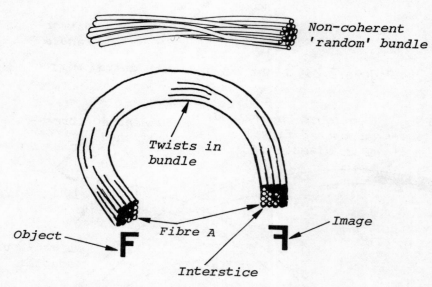

Figure 2.25 Coherent fibre arrangement

statistical connotation of the word. When even distribution of light is required, they can be truly randomly distributed.

The opposite of the random arrangement is the coherent one. Figure 2.25 shows a coherent arrangement which produces an image, as opposed to a lightguide. The image is composed of different intensities and wavelengths of light (see page 11). The manner in which these image bundles collect the image is akin to the manner in which newspaper photographs are prepared. Each fibre conveys a spot of light so that the end result is a very closely compacted raster or matrix. The image resolution depends on the diameter of the fibres. The smaller diameters of synthetic fibres have restricted uses for image guides. The majority of image guides, including those incorporated in fibrescopes, are made from glass fibres. Provided the ends are coherent the fibres can be random elsewhere.

Optical fibres, in addition to their incorporation in fibre-optic
instruments, have independent and less sophisticated uses. It would
be an exaggeration to describe them as fibre-optic instruments in
this connection. They are components. Used independently they
have three uses:

1   Transmitting or conducting light from a confined to a restricted
    place.
2   Transmitting light from an unrestricted space into a confined
    space or onto a miniature target.
3   Transmitting light from an unrestricted space to another un-
    restricted space. This use principally concerns telecommunications
    (see Section 4.5).
4   Transmitting light between two restricted spaces.

The first use is the simplest, but the second has many different
aspects. In the car headlight monitoring application, they are used to
collect light from inside the lamp and convey it to a convenient
monitoring position for the driver. The dashboard display is full
size in Figure I.2. The separate collection points can be dispersed at
the front and rear of a vehicle 1.5 X 4 m. Variations in light intensity
are not required. It is an on/off situation.

Displaying notices of variable information using a large raster or
matrix format has uses wherever people pass or meet. The written
word takes time to prepare with plate-making and the subsequent
printing processes. Incandescent lamps can be positioned as close as
fibre bundles in a raster but they do not have their intensity. Each
one requires a lamp holder and checking to ensure that it is working.
Fluorescent tubes cannot be assembled into rasters although they
can simultaneously illuminate several cut-out letters and figures.

Optical fibre bundles allow recognisable alphanumeric data to be
communicated brightly and quickly, with sufficient light to permit
magnification of the raster, from a few lamps. Figure 2.26 shows a
raster. The arrangement is described more fully in Section 4.23. The
same arrangement is used for motorway signs where the information
is not so fast moving as in the sports arena. Besides alphanumeric
display, signs and symbols can also be lit up by the multi-branch
bundles on the raster.

The second group of applications is spatially the reverse of the
above category. A lightguide can illuminate an object for inspection
by the naked eye. The lightguide may be one or more synthetic
fibres or an assembled bundle of glass fibres, possibly in a PVC jacket.
If the access is restricted so that there is no room to see round the
lightguide, the use of an endoscope or fibrescope is obligatory.

However, the threading of optical-fibre ribbon through the
confined spaces of a vehicle dashboard, so that one lamp illuminates
both instruments and legends, is spatially in an identical category.
It differs in that the light, through glow areas on the ribbon, is

distributed. The glow areas do not have the intensity of light emitted by the bundle ends in the moving-notice raster arrangement.

Very short lightguides can be used where lengths are short and cavities wider. These lightguides can be attached to incandescent, battery-powered light sources, similar to those used for taking light right to the suspected defect in a clock. Overhead lighting cannot be directed to those places so easily as a slim, optical-fibre lightguide. These lightguides avoid taking heat, always associated with the production of light, and electricity, into the inspection area. Lamp envelopes can shatter if they touch cold metal or encounter excess moisture (see Figure 2.27).

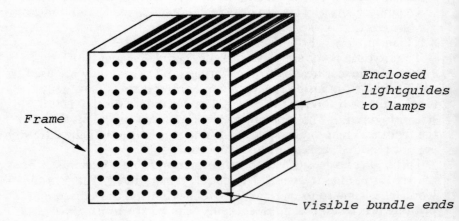

Figure 2.26 Raster or matrix

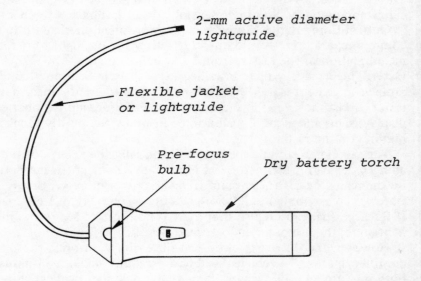

Figure 2.27 Battery lamp with small lightguide

Larger lightguides are used with larger light sources. They have the added advantage of taking more intense light than is available from any other source into the inspection area. The rigid light conduit might be used if sharp edges, e.g. in a hydraulic component, are likely to damage the jacket or if there is a need to avoid 'snaking'.

Synthetic fibres can have the *cladding peeled* to increase the illumination when inspecting components with labyrinthine tapped and drilled blind and cross-holes.

The lightguides may be up to 2 m long and 5 mm in diameter. The shapes of the holes into which they are eased varies enormously from drilled ones through extruded ones to fabricated ones in sheet metal. Synthetic fibres, 1 mm in diameter, are pushed into the long rectangular holes typically encountered when positioning assembled printed circuit boards.

Deciding when the use of a lightguide has to give way to an endoscope is sometimes difficult. A tortuous access may demand a fibrescope, if the bends cannot be seen round, or magnification of the object is required. Accesses are not, however, always curved nor magnification essential. Magnification can be misinterpreted. Where the ratio of diameter to length is less than 10, for straight holes, an endoscope is usually required. Above that ratio, problems of accessibility can be overcome in different ways. Figure 2.28 shows a light rod positioned opposite the inspector's eye and Figure 2.29 the use of an independent mirror.

Light can not only be directed onto objects but it can be used to *shine through* them when they are translucent. Quartz halogen light sources give a sufficiently intense light for this purpose. Printed

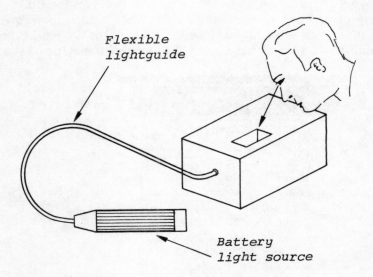

Flexible lightguide

Battery light source

Figure 2.28 Lightguide opposite the eye

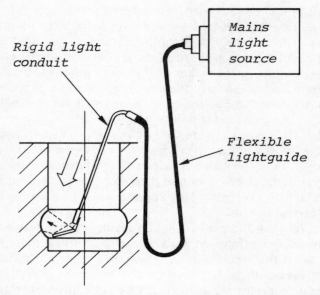

Rigid light conduit

Mains light source

Flexible lightguide

Figure 2.29 Use of an independent mirror

circuit boards can be examined for broken tracks in this, amongst other, ways.

Lightguides can have disadvantages. In dentistry they can be an encumbrance even if the light intensity in the patient's mouth is greater than that from an overhead lamp.

In the densely populated factory inspection areas, one light source can service several lightguides.

Microscope specimen illumination usually occupies the full capacity of a quartz halogen light source. When a microscope's own illumination is unable to provide sufficient, or the right type of illumination, fibre optics is used. The specimen may be opaque

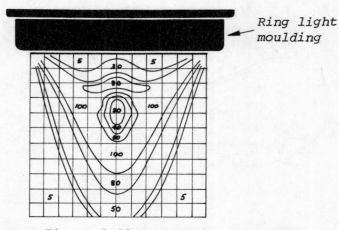

Ring light moulding

Figure 2.30 Ring light isophotes

or on a transparent slide. There are two uses of fibre optics for
microscopy—point and circular illumination. Figure 2.17 showed
a ring light. Figure 2.30 gives the isophote diagram for a ring
light using the same light source. The figures are percentages of
input light.

Figure 2.31 shows a semi-rigid swan neck, used to supplement a

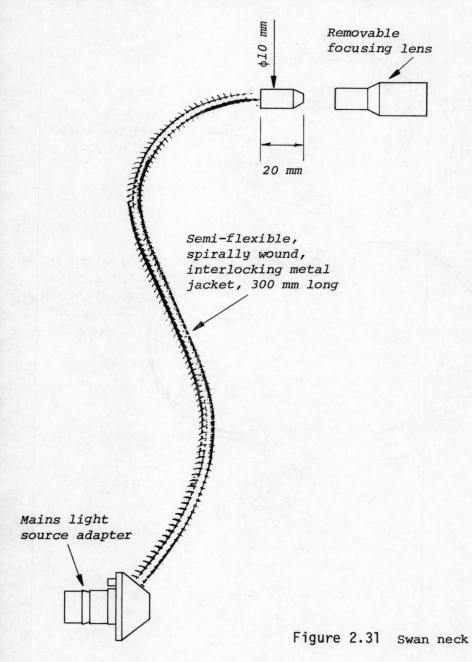

φ10 mm

Removable
focusing lens

20 mm

Semi-flexible,
spirally wound,
interlocking metal
jacket, 300 mm long

Mains light
source adapter

Figure 2.31   Swan neck

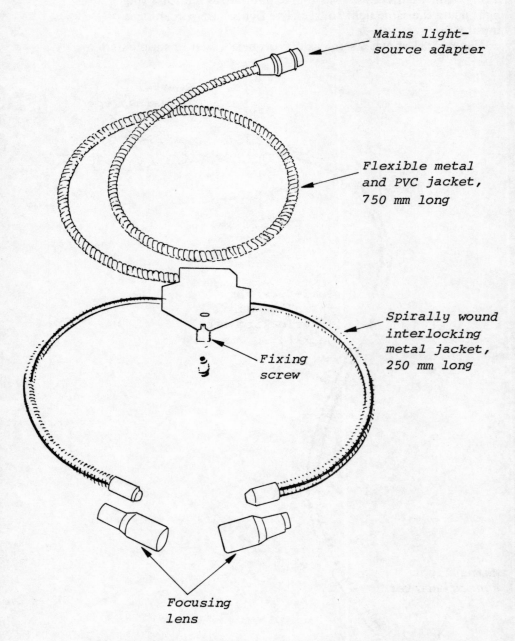

Mains light-
source adapter

Flexible metal
and PVC jacket,
750 mm long

Spirally wound
interlocking
metal jacket,
250 mm long

Fixing
screw

Focusing
lens

Figure 2.32   Double swan neck - half size

microscope's illumination, and Figure 2.32 a double swan neck. The two arms are positioned on either side of the specimen, to give shadow free light. Included in Figure 2.31 is a focusing or collimating lens. This is used to counter the effect of the fibre aperture and concentrate the light on a small area.

Ring lights have a use apart from in microscopes. They can have a magnifying lens fitted into the centre. Low-power magnification combined with intense light can be taken to large specimens which could not be placed under a microscope. In addition the ring can be positioned in many more planes than a microscope objective.

Cross-section converters have more specialised applications. They are made specially for those applications, unlike lightguides (see Table 2.3). Many of these applications are in data processing which uses marks on and holes in cards, and holes in paper tapes. The cross-section converter can either

1 Illuminate for photodiode detection, using the light barrier technique through holes.
2 By using the same technique, receive light from incandescent lamps or light-emitting diodes, or
3 By using the reflection comparison technique, illuminate and receive reflected light from marks.

All these are input-reading applications. (Fibre optics can apply a process control unit to the other form of input, magnetic tape, to supervise its manufacture, but that involves a fibre-optic instrument.)

The third case, as with all the sensing and scanning uses of fibre optics, uses coaxial fibres. Some emit light and some receive it.

The slits in Figure 2.19 correspond to the punching positions in a card.

Fibres can, if necessary, be more closely assembled than lamp and diode arrays.

Fused fibres are the last of this second large group of independent uses of fibres. To Figure 2.10 and 2.11 should be added the cone in Figure 2.33. The sides are not parallel but tapered. In some electro-

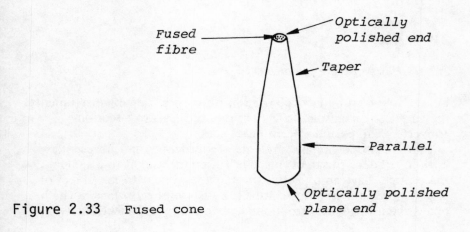

Figure 2.33   Fused cone

optic instruments, fused fibres replace lenses. They offer a number of advantages. In their flat form, when they are called faceplates, they give an image in one plane only, free from aberration. Besides being termed faceplates, which identifies their position as a component, they are called micro-channel receivers. Each short coherently arranged fibre is a channel of information. Since they need not have plane surfaces they can, with appropriate curvature, match an image to a face. They are used in low-light television cameras for image intensification. In those circumstances, every possible quantum of light flux must be collected. They can operate in half starlight conditions.

Used in cathode ray tubes, they give increased contrast, so that direct (co-planar) photographic prints can be taken. Used as front windows, they transmit the phosphorescent image to the surface of the tube.

Conventional lenses are superior to fused optical fibres in some respects. The one cannot substitute for the other. At a magnification factor of greater than 20, the outline of the fibres is superimposed on the image and obscures it.

Drawn into cones — and optically polished — fused fibres can be used to reduce or magnify. The otoscope has the reduced end in the patient's ear so that it receives ambient light from the larger end which also produces the magnified image.

The co-planar property is utilised in ultra-violet recording oscilloscopes. Here, a fused quartz fibre faceplate is used through which the beam passes onto dry photographic paper. As with cathode ray tubes this gives better resolution than would be otherwise possible — with an added advantage: the speed is greater than would be possible with a pen recorder.

Describing the use of fibres for transmitting from one unrestricted place to another for telecommunications demands an understanding of lasers and electronic devices. Moreover, this is a specialised use of fibres. A more detailed description than is possible at this point in the book has been deferred until Section 4.5.

2.3 Lenses

2.3.1 Fibre-optic applications

Lenses are not only incorporated in fibre-optic viewing instruments; they are also an adjunct to lightguides, probes, cross-section converters, lamps and electronic devices.

Fibre-optic viewing instruments — endoscopes and fibrescopes — both have a lens for an eyepiece or for connecting it to a camera. Endoscopes have an objective or object lens, next to the object or target. Inside the innermost tube is a system of relay lenses which relays the information from one lens to another. Objectives used with endoscopes are usually 30°, less frequently 45° or 10° but sometimes larger. The deep-hole endoscope has a 50° one and the modelscope a 70° objective. This figure describes the 'spread' angle of the lens. The greater the angle the larger the area viewed. The 'spread' reduces the amount of perceived detail.

Fibrescopes use coherent fibres to collect and convey the image to the eyepiece.

Attached to single and twin-armed swan necks, lenses concentrate the light on the local area of a microscope specimen.

Lenses attached to scanning probes concentrate the emitted light on a miniature target, thereby countering scatter, and collect information for the receiving fibres from the target.

A 78-mm diameter, X10 or similar, magnifying lens can be fitted into a full circle or four-point ring light so that external surface defects can be given an enlarged image.

Inside light sources an aspherical condensing lens collects light from the shadow-free parts of quartz halogen lamps. Another lens directs it to the small end-section of the attached lightguide.

Some incandescent lamps and high-radiance light-emitting diodes have a glass or resin lens integral with their envelope to produce a concentrated spot of light for the end of a probe or a slit. Large lenses are used in fibre-optic notices.

2.3.2 Types and shapes of lenses

The majority of lenses are circular, with either a plane or a curved face. The word 'lens' was originally used on account of the similarity of the first ones to the seed of the lentil plant. Although clear plastic materials are used, the stringency of the requirements usually favours optical glass for fibre-optic applications. Often there is difficulty in providing light of sufficient intensity to produce a meaningful image. Efficient transmission of the image cannot then be introduced into the optical system.

There are three forms of lens — plane, concave and convex (see Figure 2.34). The lenses can vary in diameter to suit the different sizes of fibre-optic instrument — and other optical instruments, not using optical fibres. Thickness also varies, as does curvature. There are over 300 types of optical glass — yet another variable.

The diagrams in Figure 2.34 are, as far as fibre optics is concerned, highly magnified. The smallest diameter endoscope object is only 2.7 mm in diameter.

The power of a lens is measured in *dioptres* and can have a negative or positive value. The power is usually a single or small double-figure quantity, e.g. 12.

Besides circular lenses, a recent development is the *linear echelon lens*. It acts in a line rather than circle. The lens is longitudinal and matches slits in cross-section converters dimensionally. It consists of

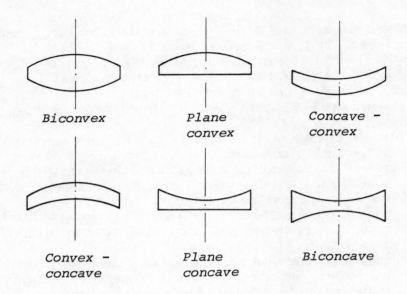

Figure 2.34   Lens shapes

a thin slice of cellulose acetate butyrate with 8 grooves per mm on one surface. Its refractive index is 1.477 and it can be used at temperatures from -4 to 80°C. Thicknesses vary from 0.5 to 2 mm (see Figure 2.35). Table 2.7 lists the focal lengths and aperture widths of these lenses; they can be used for collimation or magnification.

2.3.3   Properties and limitations

By refraction, lenses can change the direction of light so that they can magnify, concentrate and collect light. The study of lenses is vast and complicated so only two general rules are stated. Concave lenses disperse light while convex ones bring it to a focal point. The latter are the most frequently used in fibre optics.

By combining lenses in a system, images can be relayed from one focal point to another as in an endoscope (see Figure 2.36). As many as thirteen lenses may be required in the total length of an endoscope. The selection and calculation of lens values for endoscopes involves so many variables that computers are extensively used for designing optical systems.

If the optical system has been correctly designed, many of the limitations of lenses are not apparent to the user of fibre optics. Focal considerations impose dimensional restrictions on the distance between the object and the end of the fibre optic instrument, as they do between a spot of light from a lamp or light-emitting diode and the end of the fibre bundle. With endoscopes, these focal restrictions usually mean that the object has to be situated a short distance away

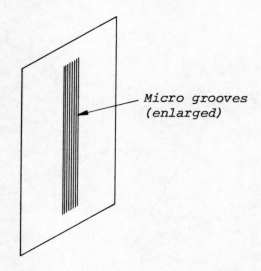

Micro grooves
(enlarged)

Figure 2.35   Linear echelon lens

TABLE 2.7

Effective focal length		Lens aperture width	
mm	inches	mm	inches
6.4	0.250	6.4	0.250
9.5	0.375	9.5	0.375
12.7	0.500	12.7	0.500
19.0	0.750	19.0	0.750
25.4	1.000	25.4	1.000

from it, e.g. 2 mm. Availability of illumination is the limitation on distant image collection.

Higher magnification, with its thicker lenses, absorbs more light because more light energy is needed to pass through the thicker glass.

Manufacturing irregularities can also impose a limitation on lenses. Air bubbles, even tiny ones, can cause diffraction.

Chromatic (or colour) aberration is attributable to white light being composed of several different colours, or wavelengths of light. Each colour is an amalgam of several different wavelengths. Due to the difference in the wavelengths of light the axes of the two extreme colours, without an achromatic lens, which is made from two different types of glass, would be detectable (see Figure 2.37). The coloured fringes which appear at the edge of some images are due to chromatic aberration.

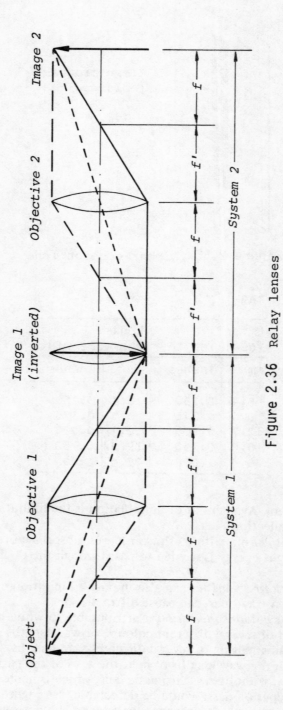

Figure 2.36 Relay lenses

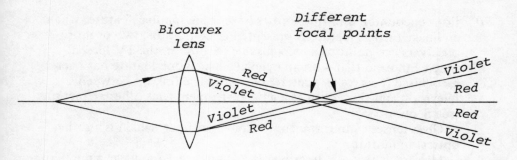

Figure 2.37 Chromatic aberration

Astigmatism in a lens, which resembles the human eye defect, prevents rays of light being brought to a camera focus when a ray strikes it obliquely. It gives the effect of two different focal lengths.

Coma is also a structural defect in lenses. A pencil of rays striking the lens at the edge is brought to subsidiary foci displaced laterally from the main circle which is not unlike the trail of a comet — hence the name of this lens limitation.

Spherical aberration occurs when parallel rays of light meet a lens or mirror. All the rays are refracted but those at the periphery of the lens are brought to a slightly different focus (see Figure 2.38).

Parallax, while not a structural limitation of a lens, is nevertheless a limitation on the use of lenses, especially eyepieces. It is the apparent change in an image caused by an actual change in the point of observation. Two observers viewing the same object through the same instrument, in the same position, without refocusing, may see a different image. Differences in the eyesight of the two observers afford only a partial explanation.

Fibre-optic manufacturers offer eye cups as an optional extra with viewing instruments. These serve to locate the eye of the observer over the eyepiece so as to minimise parallax.

Users of fibre-optic viewing instruments are connecting closed circuit television to those instruments. Since most diagnostic situa-

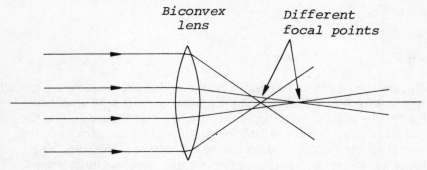

Figure 2.38 Spherical aberration

tions encountered in fibre optics have some familiar feature, when an image of that feature is presented on a monitor, two or more observers can point to a recognisable feature on the TV screen; they cannot do that with an eyepiece. Once one feature has been recognised it serves as a reference point. The link up between television and fibre optics viewing instruments has other attractions (see Section 3.5).

Glass lenses cannot transmit ultra-violet light, which is another optical limitation.

Mirrors, which are described in the next section, have some limitations similar to those of circular lenses, but parallax is not one of them.

2.4   Mirrors

2.4.1   Fibre optic applications

Mirrors have both external and internal fibre-optic uses. They are attached to light conduits to direct light from the fibres onto a small area. They are used with curved light conduits for oral inspection. In both these examples the mirror can be easily detached.

Some endoscopes are converted from direct to side view by sliding on a long revolvable tube which has a mirror at the objective end of the instrument. Different angles are available — 90° lateral, 70° retro and 115° direct oblique, which is sometimes called prograd. These mirrors reflect the light from the fibres and direct the collected light into the objective. One endoscope can then be converted to four viewing angles.

A short tube housing a lateral mirror redirects the light at the end of industrial fibrescopes.

Parabolic mirrors are used in light sources to collect light emitted by lamps and to direct it into a lightguide. Some light sources incorporate a moving mirror so that intensity can be regulated without altering the colour temperature of the light. Mirrors have been used to economise on space in data processing applications (see Section 4.7.3).

2.4.2   Types and shapes

Mirrors can be plane, convex or concave. Parabolic ones are a special concave form (see Figure 2.39). Concave mirrors tend to collect light while convex ones disperse it. Included in Figure 2.39 is the full-size mirror for converting a 3-mm diameter endoscope to the lateral viewing angle. Mirrors can be front or rear surfaced. The latter are favoured for fibre optics because they use more rugged materials.

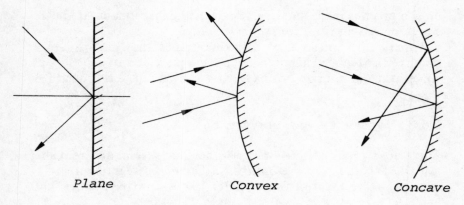

Plane   Convex   Concave

Figure 2.39 Mirror shapes

2.4.3   Properties and limitations

Mirrors redirect light by reflection. They are limited by spherical aberration and they absorb some of the light which they receive (see page 69). The object with which they are associated has to be within the focal length of a curved mirror if an acceptable image is to be collected. Curved mirrors, which are used for entertainment value, distort images so plane mirrors are almost exclusively used with fibre-optic viewing instruments where distortion must be avoided.

The other optical limitations of mirrors are the wavelengths which they will reflect. Silver produces the highest reflectivity in the visible region but aluminium is the most durable. Gold is best for infra-red light. For ultra-violet wavelengths (100-350 nm) aluminium is best. Below 100 nm platinum is used because aluminium becomes transparent.

Mirrors also have mechanical limitations. Surfaces can be scratched. A scratch introduces localised irregular reflections. Mirrors should be cleaned with optical cleaning materials. Glass and plated plastic are subject to chipping even when enclosed inside a tube, as they are when used in fibre-optic instruments.

Prisms, which are described in the next section, are free from some of the limitations of mirrors.

2.5   Prisms

2.5.1   Fibre-optic applications

Prisms are used as an optical alternative to mirrors and when a mirror cannot be used with a fibre-optic viewing instrument. If there must be no danger of a foreign body becoming loose in the object area, a prism is made an integral part of the instrument. Medical fibrescopes use prisms for lateral viewing.

When binoculars are used with endoscopes, they incorporate prisms. Right-angled adapters for the eyepiece end of endoscopes sometimes incorporate prisms for turning the image through 90°.

2.5.2 Types and shapes of prisms

There are two broad categories but there are several different shapes; not all of them have fibre-optic uses. The two categories are dispersion (see Figure 1.2) and reflection prisms (see Figures 2.40 and 2.41).

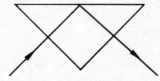

Figure 2.40   90.° Reflection prism

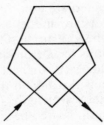

Figure 2.41   Pentagonal prism

2.5.3 Properties and limitations

Reflection prisms redirect light without magnification or aberration. Being made of glass, they do not corrode like metal mirrors.

Meridional prisms can be used for magnification. They are subject to the same mechanical limitation as glass mirrors; they can be chipped.

2.6 Filters

2.6.1 Fibre-optic applications

The different types are described by reference to their fibre-optic applications so a separate heading for the different types has been omitted in this section.

Coloured filters are used with quartz halogen light sources to eliminate, partially, those wavelengths of light which, if used for long periods during the working day, irritate the human eye.

Bright, machined-metal surfaces with their irregular reflections, are a source of irritation when the quality control of components requires intense quartz halogen light.

Colour filters, used with a similar light source and a ring light, for microscopy, improve contrast. They are used in colour photography for the same application.

Heat filters absorb dangerously high levels of infra-red light which are beyond the thermal limitations of fibres.

Black glass filters are used with ultra-violet light sources to protect the eyes of the user against permanent injury.

Interference filters allow a specified waveband to pass through them.

2.6.2   Properties of filters

Figure 2.42 shows the effect of a blue-coloured filter. The most non-blue light which is allowed to pass is 60%. The lower end of the spectrum allows only violet to pass.

Figure 2.43 shows the effect of an infra-red filter. The vertical scale of the graph is logarithmic. Heat-absorbing filters either absorb the infra-red light or reflect it back to its source.

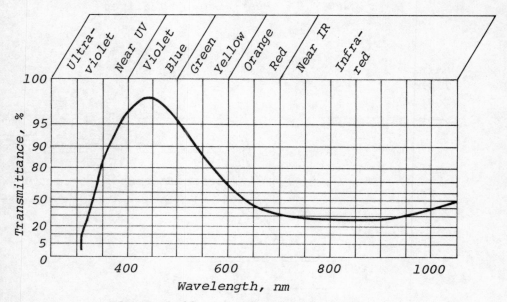

Figure 2.42   Blue colour filter

A diochroic parabolic reflector has dual filtering and light-directing functions. It removes certain wavelengths, usually the harmful infra-red ones, and reflects the useful ones. They are coated with two alternate layers of different transparent materials, e.g. magnesium fluoride and zinc sulphide. Dichroic reflectors are supplied integral with quartz halogen lamps.

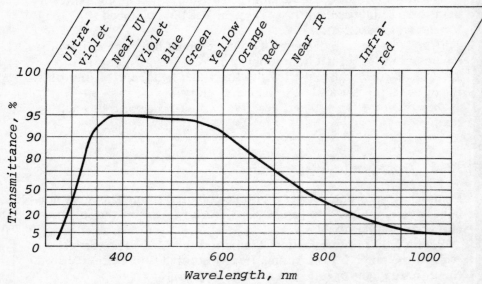

Figure 2.43   Heat-absorbing filter

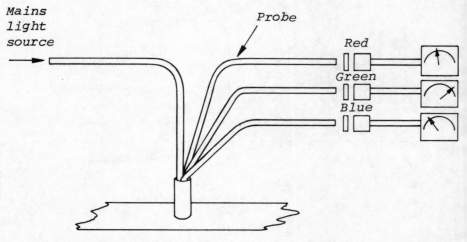

Figure 2.44   Filters selecting signal

Besides *suppressing* heat and colours, filters can *select*.
Figure 2.44 shows two coloured filters selecting different
outputs in a scanning arrangement.

2.7 Lamps

2.7.1 Fibre optic applications

Excluding ambient light, which is rarely used, there
are three sources of light for fibre optics: lamps, light-
emitting diodes and lasers (which are a hybrid).

Lamps are used to illuminate lightguides for independent
use, cross-section converters and fibre-optic viewing and
sensing instruments. There are three groups: incandescent, quartz
halogen and discharge lamps. The choice of lamp is dependent on
the following considerations:
1 The light intensity required.
2 Whether the electrical input is mains or battery (see Appendix
   3). If a portable generator is used that substitutes for a mains
   supply of electricity.
3 How much portability is required.
4 Whether the environment is hazardous and subject to the presence
   of explosive gas mixtures.
5 The space available for the lamp and the associated fibre optical
   components.
6 The position of the lamp filament relative to the end of the
   fibres.
7 The size of the bundle to be illuminated.
8 The spectral distribution of the light required — ultra-violet,
   visible or infra-red.
9 The electrical efficiency of the lamp is an occasional consideration.
10 The evenness of light distribution over the end of the fibres.
11 The colour temperature required.
12 The remedial action necessary to accommodate inherent
    limitations in the lamp.
13 The life required.

IR-light-emitting diodes are used in scanning applications and
sometimes for telecommunications. Lasers are also used for telecommunications.

2.7.2 Types of lamp

Incandescent and quartz halogen lamps have several constructional
similarities. A filament glows white hot when it is supplied with
electrical energy. Heat is always associated with light. The filament

is enclosed in a transparent envelope. Filaments are usually made from tungsten wire which has primary and scarcely visible secondary coils.

Quartz halogen lamps have a quartz (fused silica) envelope; incandescent lamps a glass one.

Figure 2.45 compares the sizes of three lamps. On the left is a 60-W lamp, in the centre a 150-W quartz halogen and on the right a 5-W, 1-W vacuum-filled incandescent lamp. The latter type is specified in terms of voltage, current, colour temperature, life expectancy, intensity at a specified wavelength and, since they incorporate a lens, spot size. Envelopes without a lens are suitable where light distribution is not critical. Spot sizes are typically 5 mm at a distance of 0.1 mm from the end of the lens.

Quartz halogen lamps have a more transparent envelope. The 'halogen' in their title is a result of adding a halogen to the filling gas. The halogens, or 'salt producers' are astatine, fluorine, chlorine,

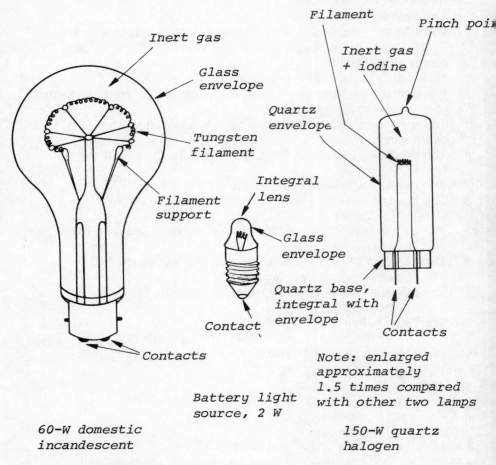

Figure 2.45 Comparison of lamp sizes

bromine and iodine. Astatine is a man-made radioactive element,
bromine is a liquid and fluorine and chlorine are gases. Iodine is a
black crystalline solid which gives these lamps their alternative
name — quartz iodine.

When iodine is heated it vaporises, forming clouds of violet
vapour; these give a coloured tint to some light sources. Inside the
envelope a reversible chemical reaction is produced with the
tungsten filament wire. There is yet another alternative name for
these lamps — tungsten halogen. The chemical reaction prevents the
tungsten being deposited on the walls of the envelope.

These quartz halogen lamps are identical with those used in ciné
projectors.

Mercury discharge lamps also have a quartz envelope and tube
because glass will not transmit the ultra-violet light which they are
used to produce. The light is formed by the recombination of ions
and atoms when electrical energy is passed through a filament (see
Figure 2.46). A stream of electrons passes through the mercury
under high pressure and ultra-violet light is given off.

2.7.3   Properties and limitations

Some of these have been mentioned, in passing, in the above section.
Quartz halogen lamps produce the most intense visible light and

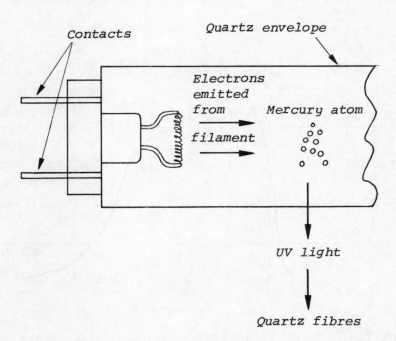

Figure 2.46 High-pressure mercury lamp (enlarged)

they are 50 per cent more efficient than incandescent lamps. Therefore they can be small in size for high output. Their load characteristics, e.g. 10 A for a 150-W light source, make their use with mains electricity supply almost obligatory.

Figure 2.47 shows the spectral distribution for a 500-W quartz halogen light source. The intensity is not even for all wavelengths of light. Spectral lamps which are larger are needed for that.

Quartz halogen lamps can be optically stabilised so that they can provide a datum for process control applications. They have two limitations. The first is the amount of infra-red light that they produce. This has to be removed by a heat-absorbing filter. Secondly, in manufacturing, the envelope is pinched at the end to seal it. If this distorted end is offered to the fibres it introduces an irregular, if, for most practical purposes, insignificant shadow. Shadow can be eliminated by mounting the lamp vertically and incorporating condensing lenses. Quartz is clearer than glass and more heat resistant. It reacts unfavourably to the secretions of the skin so care is needed when fitting these lamps into a light source.

Incandescent lamps cannot offer the same intensity. They have smaller dimensions although their intensity is so low for the very small sizes as to seriously restrict their applications. Their load characteristics are less than for incandescent lamps so they can be used with battery light sources. Small short instruments can be used with battery-powered incandescent lamps. The spot of light is

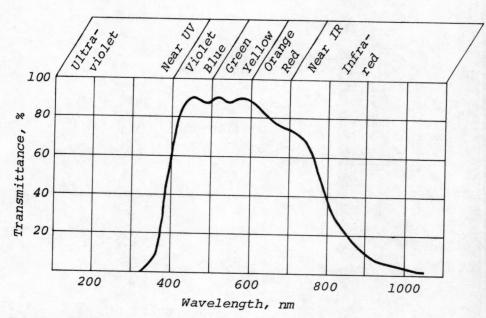

Figure 2.47 Spectral distribution for 500-W quartz halogen light source

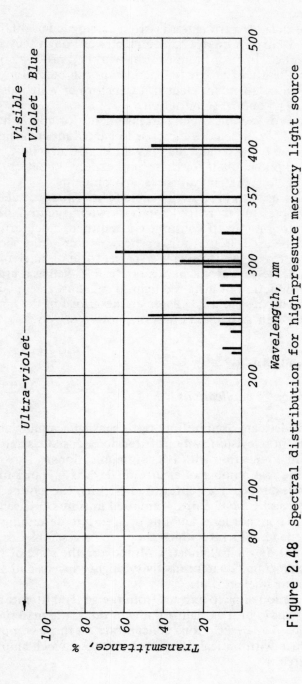

Figure 2.48 Spectral distribution for high-pressure mercury light source

adequate for projecting through the hole in some punched card readers where space is not so limited that a cross-section converter has to be used.

The life of incandescent lamps is often catalogued as 40,000 hrs but only 50 hrs for quartz halogen lamps although they last much longer in practice. For continuous, shift-working, process control applications, the life of quartz halogen lamps can be prolonged to 1000 hrs by an independent electrical transformer which also contributes to its optical stabilisation.

The fan, or convection cooling arrangement, for quartz halogen lamps, above 20 W, precludes their use in hazardous environments. Experiments are being conducted with an electrically inert, non-inflammable liquid coolant for these applications. A battery electrical supply is obligatory in many of these environments.

Lamps are also used to produce invisible light. The spectral distribution of a high-pressure mercury lamp is shown in Figure 2.48. The lines are not continuous. If continuity is required, a deuterium light source has to be used. Deuterium, or 'heavy water' is the second isotope of hydrogen. An isotope is a special form of an element. In their nuclei, deuterium atoms have a neutron as well as a proton.

For the production of infra-red light in small useable quantities, light-emitting diodes are used. These are described in the next section together with their associated opto- or photo-receivers.

2.8   Electronic devices

2.8.1   Fibre-optic applications

Electronic devices are used in fibre-optic sensing/scanning and process control units, in television systems fitted to endoscopes and fibre-scopes, and in connection with telecommunications.

Many of the vast number of electronic devices — transistors, resistors, diodes, integrators, integrated circuits, capacitors, potentio-meters, relays and C-MOS chips — are used in fibre-optic sensing instruments. There is a large amount of literature describing these devices and the circuits into which they can be assembled, and a smaller amount on their chemistry. Moreover, 'the state of the art' is constantly changing. Greater sensitivity, higher speeds and smaller sizes are routine in electronics.

This sub-section aims to extract from the general literature, much of which has audio hi-fi explantory illustrations, the material relevant to fibre optics. The descriptions concentrate on those components which interface with optical fibres, not on those which amplify the signal or switch it.

There are two types which interface with optical fibres in sensing instruments and systems — light emitters and receivers. Emitters convert electrical energy to light and receivers re-convert it.

Emitters have a function comparable to that of lamps. They are light-emitting diodes (see Figure 2.49) of the high radiance narrow-beam, infra-red variety often incorporating a lens.

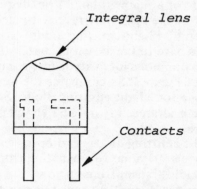

Figure 2.49 Light-emitting diode (8x magnified)

Receivers are invariably phototransistors or fibre-optic instruments, but sometimes photodiodes are used for non-standard data reading systems. Pin and avalanche photodiodes are used to receive light from telecommunication fibres.

2.8.3 Properties and limitations

The principal properties of all electronic devices, both emitters and receivers, are smallness of size, speed of response (which is less than the speed of light), longevity, high sensitivity, low electrical requirements and no moving parts. Only the electrons move. Their smallness complements fibre optics because they can be used to detect smaller

TABLE 2.8

Luminous intensity (brightness)	1.2 mcd
Peak emission wavelength	940 nm
Forward voltage	1.7 V
Continuous forward current	100 mA
Power dissipation at 25°C	200 mW
Rise and fall time	4 µsec
Storage and operating temperature	−55 to 160°C
	−131 to 212°F

targets then they could otherwise detect. The sensitivity of a phototransistor when coupled to an optical-fibre probe is such that a 0.05-mm line or a contrasting background can be detected 4 mm away at a speed of 500 Hz. Beyond that distance a lens is needed to counter the effect of the fibre aperture angle and absorption of the light by air. Infra-red light is practically immune to disturbance by ambient light.

Light-emitting diodes are used when the 40,000-hr catalogue-specified life of a filament lamp is inadequate. They convert about 60 per cent of the light they receive to electricity. Table 2.8 gives the specification of a light-emitting diode.

Receivers have to be chosen to match the peak emission wavelength. The luminous intensity is low compared to a quartz halogen light source! Figure 2.50 compares the luminous intensity with the viewing angle for a light-emitting diode. Sensitivity decreases rapidly as the angle is altered. Figure 2.51 shows the effect of ambient temperature.

Both light-emitting diodes and opto-receivers can be adversely affected by electrical noise emanating from nearby large electric motors in radio transmitting stations.

Phototransistors are light-sensitive resistors enclosed in a transparent envelope. They have high electrical resistance in the dark but when exposed to light the resistance reduces approximately inversely to the illumination (see Figure 2.52). Cadmium selenide may be the active layer in a photofinish but other semiconducting materials are also used.

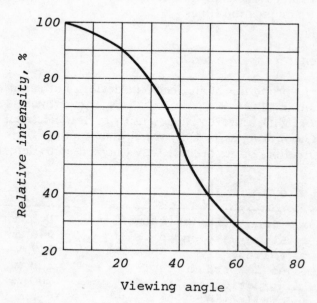

Figure 2.50 Relative intensity versus viewing angle

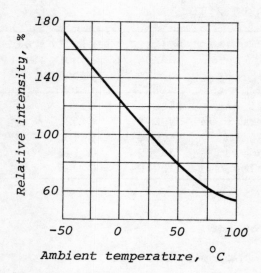

Figure 2.51 Luminous intensity versus temperature for phototransistor

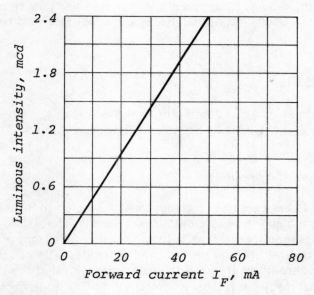

Figure 2.52 Resistivity of phototransistor

Different materials for these devices have different characteristics and properties (see Table 2.9). Silicon has better response to visible light than gallium arsenide phosphide and better conductivity than germanium. However availability of some of these semiconducting materials constitutes a problem. Semiconducting materials are also used in telecommunications lasers (see next section).

TABLE 2.9

Material	Melting point, ^{o}C	Electrical conductivity, (mho/cm) x 10^{6}
Silicon	1420.0	0.100
Germanium	958.5	0.022
Gallium	1600.0	0.580
Cadmium	320.9	0.146
Selenium	220.0	0.080

2.9 Low-power lasers

2.9.1 Fibre-optic applications

Semiconductor low-power infra-red lasers are used to launch the light into the optic fibres for many telecommunications applications. They have a much faster response time and a narrower beam than light-emitting diodes. High-power lasers are not used for fibre-optic telecommunications but optical fibres are sometimes used to transmit them.

2.9.2 Definition

The word 'laser' is formed from the initial letters of the word which define it – Light Amplified Stimulated Emission Radiation.

2.9.3 Operation principles

In the first lasers a crystal, e.g. ruby, was subjected to powerful flashing by a high intensity lamp, usually xenon. The size of these xenon lamps inhibits their use with the small end-sections of assembled optical fibres in probes and lightguides. The xenon lamp excites the chromium atoms in the ruby crystal so that the electrons jump further away from their nucleus. When they jump back again they emit light. The crystal has silvered ends which reflect the light inside the laser medium so that more atoms are raised to the excited state. In the excited state, light of the same wavelength, phase and direction as the stimulating light is emitted by the laser medium (see Figure 2.53).

Low-power gas lasers, using helium, neon and (for fibre-optics telecommunications purposes) semiconducting materials, are being developed and used.

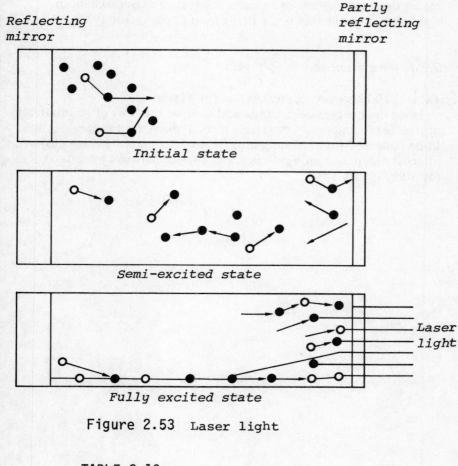

Figure 2.53 Laser light

TABLE 2.10

Wavelength range	800-900 nm
Pulse width	10 nsec - 5 μsec
Repetition rate	5×10^{14}
Beam:	
Width	1 - 1.5 mm
Divergence angle	15°
Output power	15 to 10 mW

In a semiconducting laser, the planar *pn* junction in a light-emitting diode is bounded by two parallel reflecting surfaces. Laser action takes place when free electrons in the conductor band are stimulated to recombine with holes in the valence band. In recombining, the electrons give up energy which is radiated as light quantums. Gallium arsenide (known as a direct-gap laser) amplifies better than germanium and silicon indirect-gap lasers. Gallium aluminium arsenide lasers are being used experimentally.

2.9.4 Properties and limitations

Table 2.10 lists some typical values for a laser.

Laser light is monochromatic and follows the laws of geometrical optics. Semiconductor lasers have a small diameter beam which has to be collected for launching into fibres. If operated at high current, internal dissipation of light occurs so they are always operated on low duty cycles.

Three

FIBRE-OPTIC INSTRUMENTS

3.1 Résumé

All fibre instruments incorporate one or more of the fibre-optic
components described in Chapter 2. Table 3.1 lists the components
and instruments. Telecommunications is a system rather than an
instrument so it does not appear in the list. Mechanical and electrical
components and sub-assemblies are also omitted from the table.
Quartz halogen light sources sometimes use electric fans for cooling,
endoscopes incorporate stainless steel tubes and process/control
units have sheet metal sub-assemblies.

3.2 Light sources

3.2.1 Associated equipment

Table 3.2 lists the equipment associated with light sources.
  Quartz halogen light sources are the most widely used and most
versatile but, in certain situations (see page 78) incandescent lamps
are preferred. High-pressure mercury light sources are used for the
provision of ultra-violet light which can only be transmitted by
quartz fibres and, to a much lesser extent, by some synthetic ones.

3.2.2 Design considerations

Any *excess of heat* has to be dissipated so a fan is incorporated
which draws or blows air across the quartz halogen lamp. In dual-

TABLE 3.1

Instrument	Components	
	Always	Sometimes
Light sources	Lamps	Lenses Filters Mirrors
Endoscopes	Optical fibres Lenses	Mirrors Prisms Lamps
Fibrescopes	Optical fibres Lenses	Mirrors Prisms
Scanning units·	Optical fibres Electronic devices	Lenses
Process control units	Optical fibres Electronic devices Lamps	Filters Lenses

voltage light sources, with their bulky transformers, the mass of metal is large enough to permit convection cooling. The portability of the light suffers if they are offered for both 110 V and 220/250 V; 20-W light sources can be cooled by convection. Experiments are taking place using inert liquid coolants but the refractive index of the coolant has to match that of the fires whilst having a specific gravity high enough to absorb the heat. The latter difficulty may restrict their use to the lower energy lamps in hazardous environments as an alternative to lower output incandescent lamps. Heat-absorbing filters which have a smoky glass appearance, are incorporated to remove excessive infra-red hot light. Often quartz halogen lamps also incorporate an anti-UV mirror.

Lamp life also affects the design considerations of quartz halogen light sources, Process control units which are fixed in one situation have an independent transformer incorporated in the chassis. The majority of light sources have a transformer of a lower grade inside their metal case, so that standard ciné projector lamps can be used. Long-life light sources not infrequently use airport runway landing lamps which have a life expectancy of about 2000 hr.

The unpredictability of lamp life has resulted in light sources intended for medical use being duplicated. The medical viewing instrument is only momentarily deprived of light in the event of a lamp failure. A Y-shaped lightguide connects onto the two lamps which can be operated by a three-position switch or switched by an electro-optic arrangement. Earlier medical light sources had a single lightguide with a slide arrangement for moving the good lamp across to the end of the lightguide to replace the failed one.

The *light intensity* may require regulation. There are applications where it is too intense. In addition to colour filters, a stepless regulation can be achieved electrically, optically or mechanically.

A rheostat can be used to reduce the electrical energy being conducted to the filament, but this affects the colour temperature of the light. At the lower input voltages the light has the yellowish

TABLE 3.2

Type of light source	Assembled optical fibre configuration	Fibre-optic instruments
Quartz halogen - mains	Single- and multi-branch lightguides up to 13-mm input diameter Ring lights Single and twin swan necks Cross-section convertors	Endoscopes and fibrescopes Process control units
Incandescent - battery - mains	Single lightguides up to 4.5-mm diameter	Short/medium diameter endo-scopes, e.g. 4.6 x 250 mm Similar fibrescopes Scanning units - visible light
High-pressure mercury - mains	Quartz fibre light-guides up to 4-mm diameter	Quartz fibre endoscopes, e.g. 5 x 150 mm
Light-emitting diodes - mains	Single, small-section lightguide or probe, e.g. 2φ x 500 mm	Scanning units

Figure 3.1   Iris diaphragm

colour associated with an under-run filament. Electrical regulation permits the running up and running down of lamps although the practice is of unproven value.

Optically, intensity may be steplessly varied by a moving mirror intermediate between the quartz halogen lamp and the end of the fibre bundle. This method does not affect colour temperature. Mechanical variation of intensity, through an iris diaphragm, similarly does not affect the colour temperature because the output from the lamp is unaltered. Iris diaphragms (see Figure 3.1) are similar to camera shutters.

Table 3.3 gives some light specifications. The intensity of the 20-W model does not justify the incorporation of intensity regulation.

The *weight* of the 100-W light source is greater than that of the others because an independent transformer is used. This is a process control light source where colour temperature is not of paramount importance. The 500-W light source produces less light than the small 150-W lamp because it incorporates condensing lenses to produce coherent light.

All the quartz halogen lamps can be over-run to improve on light intensity, but it adversely affects the life of the lamp.

Similar details are not available for light sources which incorporate an incandescent lamp and use batteries. The discharge characteristics of the batteries are such that the data would be very difficult to obtain. They weigh above 250 g and their light output is about 10^3 lx for 3-W, 5-V lamps. When lead acid accumulators are used with 6-V, 20-W lamps, there is some improvement on output, at the expense of an increase in weight.

3.3   Endoscopes

3.3.1   Applications

Endoscopes are used for illuminated, magnified viewing of objects situated in cavities where there is an in-line access large enough for the instrument to enter. The deep-hole endoscope is a special type which does not enter the cavity. Some holes, although not straight, but with a large radius of curvature and tapering cavities which narrow to dimensions smaller than the endoscope, can be viewed with their aid. They are used when no heat or electricity must be present in the object-viewing or target area. The lamp is inside the light source in the unrestricted space.

TABLE 3.3

Rating W	Power consumption W	Lamp V/W	Input V	Weight kg	Output lx	Lightguide max.diam. mm	Colour temperature K
20	25		220/240 or 110/130	1.1	7×10^5	4	3400
100	150	16/100	Ditto	6.2	5.4×10^4	13	2400
150	180	15/150	Ditto	4.5	2×10^6	13	3400
250	300	24/250	Ditto	4.6	2.5×10^6	13	3500
500	550	220/500	Ditto	3.2	1.5×10^6	13	3200

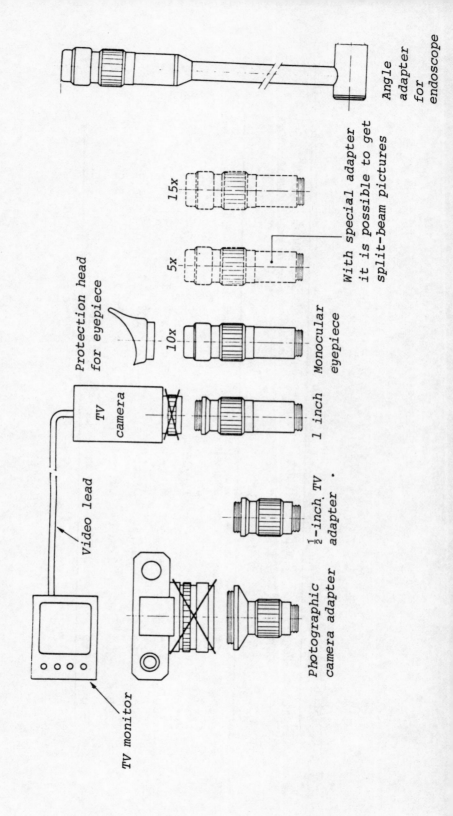

Angle adapter for endoscope

With special adapter it is possible to get split-beam pictures

15x

5x

Protection head for eyepiece

10x

Monocular eyepiece

TV camera

1 inch

Video lead

½-inch TV adapter .

TV monitor

Photographic camera adapter

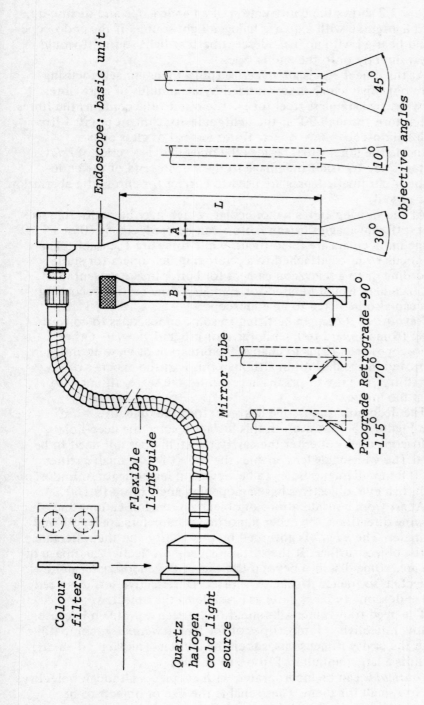

Figure 3.2 Endoscope construction and layout

Figure 3.2 shows the construction of an endoscope and its link-up, via a lightguide with a quartz halogen light source. If the endoscope could be used with an incandescent battery light source it would screw directly onto the endoscope.

At the object end is an *objective* (object lens) or self-focusing fibre objective which is surrounded by an annulus of glass fibres between two stainless steel tubes. The outer tube contains the fibres which turn through 90° at the lightguide attachment point. Ultra-violet endoscopes have quartz fibres instead of glass ones.

Inside the inner tube, besides the objective, is a series of *relay lenses.* They transmit the image to the viewing end of the endo-scope. Achromatic lenses are used to correct for chromatic aberration (see page 8).

At the viewing end is a monocular, which may be removable so that different magnification factors can be used, e.g. 5, 10 or 15. Some monoculars are fixed-focus, while some are focusable. The monocular can be attached to a photographic camera for static recording or to a television camera for better image presentation. A video camera would be used for animated slow-motion recording; this can also be fitted to an endoscope.

Revolvable tubes can be fitted to some endoscopes to convert them from forward to lateral, retro or prograd viewing. Other endoscopes use prisms to change the direction of viewing and gain improved image definition although this is at the expense of versatility. The use of prisms necessitates the segregation of fibres from the image.

The deep-hole endoscope has an extra wide objective − 50° (see Figure 3.3 opposite). In this instance, since the deep-hole endoscope does not enter the cavity the stem does not need to be long. The wide-angle lens enables the walls of very small cavities, e.g. 0.8 mm diameter holes to be viewed in magnification. Endoscopes with 'fish eye' objectives have an opening angle of up to 160°.

Apart from considerations of objectives, magnifications and viewing directions, the other important parameters are length and diameter. The *length* is governed by portability and the reflectance of the object surface. If the surface is highly reflective, an image of the area some distance beyond the end of the endoscope can be collected, so the length can be shorter. Unreflective surfaces need the endoscope to be as close as possible to the object.

The need to maximise illumination is often a problem because of low reflectivity. Endoscopes of the largest *diameter,* compatible with the cavity dimensions, should be chosen. The larger diameter permits a larger annulus of fibres to be used.

Graticules can be incorporated with eyepieces although objectives are too small for them. These enable the size of objects to be estimated on a comparability scale. Known values have to be suited

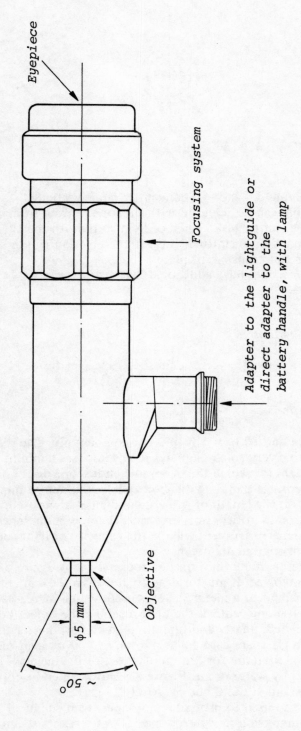

Figure 3.3

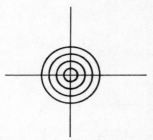

Figure 3.4   Graticule - half size

to the scale. Figure 3.4 shows a concentric circle graticule.

Beam splitting can be achieved with the lowest order of magnification. This is the splitting of a beam of light so that it is partly reflected and partly transmitted. It is effected by adding non-metallic dielectric films to a transparent plate.

Medical endoscopes have additional features — these are described in Section 4.14.

3.3.3   Specifications

Table 3.4 lists a typical range of endoscope specifications.

3.3.4   Limitations

Endoscopes are limited optically and environmentally. The difficulty of manufacturing very small diameter lenses imposes a minimum diameter on them for which the deep-hole endoscope does not entirely compensate. Handling difficulties in manufacture impose an upper limit on the length of these small diameter endoscopes. Light attenuation in optical fibres restricts the maximum length of the larger diameter instruments, while the capacity of light sources restricts their maximum diameter.

Their environmental limitations are thermal and pressure ones. Where heat is involved, it must not be greater than the cement of the fibres can withstand. If a metal jacket is used, which increases the effective diameter and reduces the light reaching the object because a window is needed, quartz and sapphire, which have better thermal properties than glass, are used for windows, and either heat-resisting stainless steel or titanium for the tubular part of the jacket. The jacket is cooled by water or air. Figure 3.5 shows an endoscope in a jacket to withstand 1200°C and 50 Atu (725 psi).

Jackets are also used to protect endoscopes from chemical attack. For constant immersion in water, a special water-resistant cement has to be used for the fibres surrounding the endoscope objective.

TABLE 3.4

Diameter		Length, mm	Objectives		
Without revolvable tube,mm	With revolvable tube,mm		10°	30°	45°
3.0	3.2	60		x	
		125		x	
		185		x	
4.6	4.9	60	x	x	x
		125	x	x	x
		190	x	x	x
		250	x	x	x
		500	x	x	x
7.0	7.4	350	x	x	x
		700	x	x	x
9.6	10.0	125	x	x	x
		250	x	x	x
		500	x	x	x
		750	x	x	x
		1000	x	x	x
		1500	x	x	x
		2000	x	x	x
14.5	15.0	250	x	x	x
		500	x	x	x
		750	x	x	x
		1000	x	x	x
		1500	x	x	x
		2000	x	x	x

Magnification - x5, x10, x15
Viewing directions - forward, lateral, retro and prcgrad

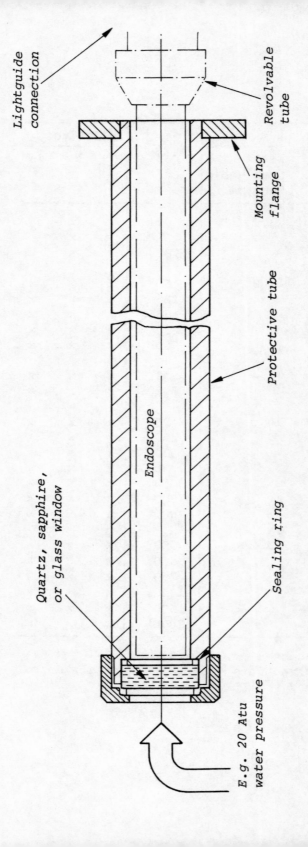

Lightguide connection

Revolvable tube

Mounting flange

Protective tube

Endoscope

Quartz, sapphire, or glass window

Sealing ring

E.g. 20 Atu water pressure

Figure 3.5   Endoscope in protective jacket

The limitations of endoscopes are such that they are closely related
to other members in the optical instrument genealogical table —
intrascopes, contact endoscopes, microscopes and modelscopes. With
the occasional exception of modelscopes, these instruments do not
incorporate optical fibres. They are, however, often confused with
fibre-optic viewing instruments.

Figure 3.6 shows an intrascope. The lamp is next to the objective.
This eliminates light attenuation because the light does not have to do
a double journey — once through the fibres and return through the
lenses. But there is electricity and heat in the object area. Table 3.5
shows how intrascopes take over from endoscopes for the larger
lengths and larger diameters. Technoscopes are a brand name for
intrascopes.

For portability, intrascopes are made in sections which can be
assembled to make up a longer length. They are available with several
interchangeable viewing directions, as well as angled monocular and
binocular eyepieces and camera adapters.

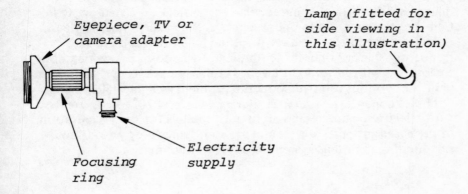

Figure 3.6 Intrascope

TABLE 3.5

Outside diameter		Basic length		Maximum useful length	
mm	inches	mm	inches	mm	inches
9.5	0.374	1,750	70	2,800	110
13.0	0.512	2,500	100	4,500	177
19.0	0.748	2,260	90	7,860	309
26.0	1.024	1,300	51	13,000	512
34.0	1.339	1,300	51	16,300	642

Contact endoscopes use ambient light, since they do not penetrate deep into cavities. The surface can actually be touched by the instrument. These are low-power instruments 'which use ambient light to collect an image from the surface of materials such as liquids, pastes and powders, which could not be conveniently examined by a microscope. Diameters range from 2 to 30 mm and lengths from 25 to 500 mm.

Optical microscopes are available in many different forms. They do not enter cavities when they are used to view them. Their direction of view is forward. Illumination is apart from the lens. They are bench mounted. The image can be presented through monocular, binocular or trinocular eyepieces or projected onto a screen. The range of magnification is large with small increments. Cameras can be fitted.

Endoscopes usually enter cavities. They can have four viewing directions. Illumination surrounds the lens. They are portable, although they can be mounted on stands which have drive units and safety stop collars similar to those used with microscopes. The latter device prevents an objective in the fibre ends being damaged by physical contact with a sharp object. Endoscopes present their image usually through a monocular. Magnification values are low. Cameras can be fitted to endoscopes, as they can to microscopes.

Microscope images, as a result of the larger lenses used, are frequently sharper than those obtained by endoscopes, which have to contend with lower light levels. If microscopes have insufficient light, fibre optics can overcome the problem (see page 60).

Modelscopes have a very wide-angle lens, e.g. 70°. They are used with television cameras and architects' models for the visualisation of proposed property development. Sometimes they have fibres surrounding their lens system to improve illumination.

3.4   Fibrescopes

3.4.1   Applications

Fibrescopes are used for magnified, illuminated viewing when there is no in-line access between the eyepiece and the object (see Figure 3.7). They can negotiate multiple curves with small radii. Their alternative names are flexiscopes and imagescopes. Uniscopes are fibrescopes with a semi-flexible sheath which increases their bending radii. Gastroscopes, colonscopes and bronchioscopes are fibrescopes used for the examination of the particular parts of the anatomy suggested by their names. Needlescopes are 2.5 mm diameter fibrescopes; they are up to 250 mm long.

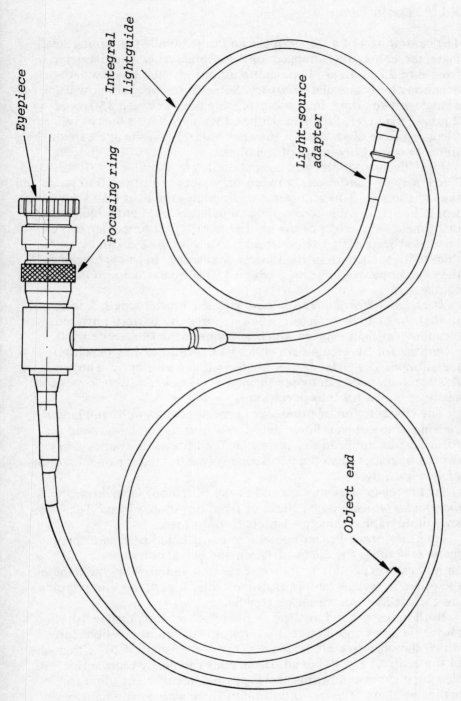

Figure 3.7 Fibrescope

Eyepiece

Focusing ring

Integral lightguide

Light-source adapter

Object end

The central part of a fibrescope is an image bundle containing small diameter, coherently arranged, optical fibres, ranging in diameter from 8 to 25 μ). Image bundles are usually circular but they are occasionally rectangular or square. Some fibrescopes are now using a single self-focusing, image-transmitting fibre between 150 μ and 2 mm in diameter. Like endoscopes, they may have a lens or self-focussing fibre objective but they can also rely on the aperture of the fibres to collect the image of the object.

Partially surrounding the image bundle are *illumination fibres.* These may be continuous between the object end of the fibrescope and the light source. This arrangement eliminates the light losses which would be attributable to coupling an independent lightguide to the instrument. The effect of the arrangement is that longer lengths can be made because total attenuation is less. Fibrescopes can be more than twice the length of the longest endoscope. In this arrangement the fibrescope would be used with a 150-W quartz halogen light source.

Detachable lightguides are used with the shorter length fibrescopes so that they can be replaced, when required, by battery-powered incandescent light sources which screw onto the fibrescope itself.

Jackets for fibrescopes are either PVC, semi-flexible, or woven metal braid. The latter can be treated with an antiseptic. The semi-flexible swan neck can be pre-formed. PVC-jacketed fibrescopes can easily be coiled for transportation.

The construction of fibrescopes provides for *forward* and *lateral* viewing of the object. Short detachable mirrors can be screwed on for industrial applications; prisms are used for medical ones. They cannot be removed so the fibrescope is constructed for one direction of viewing only.

At the image presentation end of the fibrescope is an *integral monocular* which may be either of fixed or variable focus. Television and photographic camera adapters fit onto these.

Medical fibrescopes are sometimes constructed with *additional features* – tubes for the insufflation and extraction of gases, such as air and carbon dioxide in the one direction and possibly methane in the other, as well as liquids. Bowden cables for cutting and gripping are enclosed within the jacket (see Section 4.14).

Both industrial and medical fibrescopes are being made with a moveable distal end. Figure 3.8 shows the different positions into which the objective can be moved. These are usually 120° either side of the centre, i.e. 240° in all. Distal ends can partly replace side viewing mirrors and prisms provided there is sufficient room to manoevre them. The incorporation of these additional facilities increases the diameter of fibrescopes.

Figure 3.8   Distal end

3.4.3   Specifications

Table 3.6 gives the optical and dimensional specifications for a range of industrial fibrescopes. Not included in the table are thin (2.2 mm diameter) and thick (25 mm diameter) ones which have been specifically made. There are hardly sufficient fibres in the 2.2 mm diameter bundle (its outside diameter) to collect and convey a meaningful image. Its ability to do this would be further reduced should there be any fibre fractures.

Poorer image resolutions can be tolerated for larger diameters in industrial applications but many medical fibrescopes offer a very good resolution of 50 line pairs/mm.

3.4.4   Limitations

Although Table 3.6 shows an infinite focal length beyond a minimum distance from the object, the availability of light restricts the distance of the fibrescope objective from the object.

'Snaking' on the way to the object area often presents a problem. Difficulty in countering this snaking may give rise to uncertainty as to whether the whole of an object area has been viewed.

The jacket restricts the *bending radius* of fibrescopes. A swan-neck jacket increases it by approximately 50 per cent over PVC-jacketed fibrescopes of similar diameter. The softer, more flexible, jacketed fibrescopes are more prone to damage than those jacketed in spiral or woven metal.

Magnification is restricted to the lower orders, e.g. ✕ 20, because the outline of the fibres can become superimposed on the image. This limitation applies when complementary instruments such as those described in the next section are used. If the image is presented on a television screen, the construction of the camera tube and that of the fibrescope can give it a mosaic appearance.

Diameter w/out mirror, mm	3	6	8	10
Diameter with mirror, mm	-	7.5	9.5	11
Working lengths (not overall), mm	830 1000 1300	870 1000 1300	870 1000 1300	870 1000 1300 1800 2300 2800 3800 4800
Bending radius, mm	35	40	40	40
Objective angle, degrees	24	40	40	40
Image resolution, line pairs/mm	37	31	28	25
Focal length, mm	7- ∞			

3.5 Complementary instruments

3.5.1 Types

There are two types of complementary instrument, used with fibre-optic viewing instruments; closed-circuit television cameras and photographic cameras.

Television cameras and monitors can be on-line, displaying information simultaneous with the viewing of the object, or they can record the information on videotape. With few exceptions, television systems are in black and white since colour demands a much higher light level. This can be a severe constraint when all the available light has to be emitted by an annulus of optical fibres at the end of a fibre-optic viewing instrument.

Photographic cameras are still or ciné. The latter is rarely used because the use of videotape enables the recorder to monitor what is being recorded, whereas the ciné camera does not. Cameras can either produce a negative or an instant picture.

Negatives are supplied in roll form. After exposure they have to be developed and printed. Instant pictures are supplied in cut sizes. Development and printing takes place inside the camera.

Since quartz halogen light sources emitting light at colour temperatures of 3200 and 3400 K are available, photographic cameras can be used for colour purposes. Artificial light colour film is balanced for one or other of these colour temperatures.

In this section, photography and television complement fibre optics. In Section 4.20 the roles are reversed. Fibre optics complements photography.

3.5.2 Fibre-optic applications

Both photographic and television cameras are used with endoscopes and fibrescopes. Television cameras are used in the following circumstances:

1   To make an image available to more than one person. Disputed or borderline inspection and diagnostic situations call for a second opinion. Supervisors need to be able to point visually to a feature whereas if only an eyepiece is used they are restricted to oral explanations.

2   To present the image in a more convenient form for high volume situations. Viewing a screen is more convenient than peering through a monocular or binocular for the duration of the working day; even if the light source has colour filters and intensity regulation to provide a measure of variation.

3   To achieve greater magnification than is possible with an endoscope. Monoculars do not usually exceed a magnification of $\times 15$, whereas $\times 700$ has been achieved with a television.

4   To achieve better resolution than is possible with a small diameter television camera, e.g. 19×305 mm. Their image resolution is 350 lines whereas broadcast television is 625, and that is inferior to some closed-circuit systems.

5   To retrieve information from a cavity too small to be entered by a camera. The smallest diameter fibre-optic viewing instruments can enter cavities as small as 3.2 mm diameter and 185 mm long.

6   To retrieve information from cavities where the temperature range is unacceptable to television cameras. The range for television cameras is -10 to 50°C but for endoscopes -4 to 120°C and -20 to 70°C for fibrescopes without heat-protective jackets.

7   To quantify a scan area for dimensional checking of a feature. An electronic counter counts the number of spots of a particular colour so that a comparator can judge whether the feature is inside or outside a dimensional limit.

8   To record information from cavities in environments where handling of fibre-optic equipment is difficult. Plant inspections for insurance purposes may need to be evaluated by more than one specialist but, on site, access to an inspection point is difficult even for one person.

9   To make a recording when it is not possible to take photographs.

10   To make an animated recording; possibly for slow-motion analysis.

11   To overcome inadequate light availability.

Photographic cameras are used:

1   To avoid the capital costs of closed-circuit television.
2   To provide a convenient form of record for filing and easy retrieval.
3   To record colour differences in the object area.

3.5.3   Construction of the link up

The means by which cameras are attached to endoscopes are shown diagrammatically in Figure 3.2. Sometimes the monocular is removable so the camera adapter replaces it. Other endoscopes, like fibrescopes, have a non-removable monocular so the adapter fits over the monocular.

The emphasis in this sub-section falls on the television link-up because popular technical magazines and its longer history make information on photography more readily available. Moreover, since reference has been made to the use of fused fibres in electro-optical systems (page 63ff) the opportunity is taken to describe those uses more fully.

A television camera converts the optical image received from the fibre-optic viewing instrument into and electronic one. Figure 3.9 shows a camera which might be used to complement a fibre-optic instrument. It would be much lighter in weight than those occasionally glimpsed during broadcast television presentations, e.g. 4 kg (9 lb). The outer case might be circular or rectangular. Either can be held in the hand.

The *front lens* of the camera tube could be a fused faceplate acting as an image intensifier. The diameter of the tube, 15 or 25 mm, requires a matching adapter for linking to the fibre-optic instrument.

The screen or *mosaic* in the camera tube is covered on the one side with tiny dots of caesium. The dots are so close together that the mosaic is indistinguishable from a uniformly coated surface. Each dot is photosensitive — a miniature photo-electric cell.

Caesium is one of the alkali metals along with lithium, sodium, rubidium and the radio-active element, francium. They all dissolve in water with a violence that ranges from vigorous to explosive.

As the caesium receives light from the last lens in the fibre-optic system or instrument, electrons, proportionate to the intensity of the light, are released. Intense light releases the most electrons. The variation in light intensity has to be detected if the ultimate electronic image is to have tone.

Detection is effected by *scanning* the uncoated side of the screen with a narrow beam. This scanning, or sweeping, releases electrical charges from the caesium. For a 625-line system the mosaic is usually scanned at 30 Hz. Each complete scan is called a frame. The

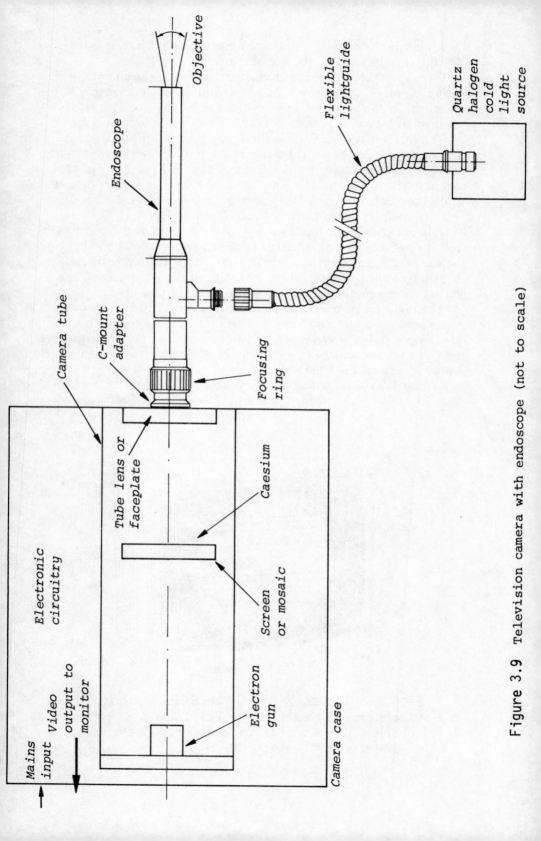

Figure 3.9 Television camera with endoscope (not to scale)

beam inside the camera is deflected magnetically by coils — one for the vertical and one for the horizontal.

Corresponding to the *electron scanning gun* in the camera, which scans the mosaic, and synchronised with it, is an electron gun in the monitor. In closed-circuit television the camera and monitor are connected by a video lead.

The *receiver* is basically a cathode ray tube (CRT) which may incorporate fused fibres at its narrowest point. It works with light waves of video frequency. The front of the CRT has a fluorescent coating which gives off specks of light when electrons are 'shot' at it by the electron gun. A popular receiver or monitor size for use with fibre-optic instruments is the 230-mm monitor.

Fluorescence is the name given to the emission of radiation as the result of absorption of radiation of shorter wavelength, e.g. light. Ultra-violet light, which requires quartz fibres, is frequently used for fluorescence. Measuring the intensity of the fluorescence and its nature is a method used for evaluation in biochemistry.

The majority of television systems scan the number of lines on the screen on a scan one, miss one, or on a random basis. This scanning is called *interlacing*. Persistence of vision in the human eye prevents the movement of the scanners being detected so a complete image is seen all the time.

Colour television, which may have an unrealised potential for

Figure 3.10 Image on television monitor. The illustration shows the bands inside a 6-mm internal diameter fluid seal viewed through a side-viewing endoscope

medical applications when linked to fibrescopes, uses the fact that white light is an amalgam of colours. The primary colours − red, blue and green − mixed in different proportions, can produce all other colours. Colour television cameras are three cameras in one tube with optical filters to split the light into its three component colours.

The monitor CRT has three fluorescent coatings to correspond to the three electron guns in the camera.

Photographic cameras work on the principle that light initiates a conversion process, when used with certain light-sensitive chemicals. Whereas the lens of television cameras is removed to be replaced by an adapter, with a C-mount, for attaching the fibre-optic instruments, the photographic camera lens is undisturbed. Some instant-photograph cameras are made specifically for attaching to scientific instruments, of which fibre-optic instruments are only one genre. A photographic film can accept between 10^4 and 10^6 channels of information per square centimetre. Moreover, light scatter in the chemical coating of the film results in the channels blending into one another.

Focusing of the link-up is effected by a focusing ring on the adapter.

3.5.4   Limitations

Availability of light of sufficient intensity may be a potential limitation but complementary instruments enable this to be overcome. Low-light television cameras with silicon intensified target tubes, and possibly incorporating image intensifiers, replacing the more usual vidicons, have almost eliminated the problem. Unless these sophisticated tubes are used, in the circumstances where objective and object are distant or the object is of poor reflectivity, light level will be a limitation on the use of a television link-up.

Photographic cameras, which are used for stills, can overcome light levels by longer exposure times. Infra-red photography also has potential here provided the light emitted is compatible with the thermal limitations of the optical fibres inside the fibre-optic viewing instrument.

Limitations on the extent of magnification for fibrescopes have been described elsewhere (see page 103). Some television cameras 'bloom' if subjected to unacceptable levels of infra-red light.

Data on exposure times for photographic cameras linked to fibre-optic viewing instruments is scarce, as are photometers for this specialised application. Photographers, in the circumstances, tend to proceed empirically.

The image obtained through a fibre-optic instrument is usually circular. Since photographic prints and television monitors are rectangular, not all the available space is used (see Figure 3.10).

3.6.1 Applications

Scanning units are used to detect, without human intervention or physical contact, miniature targets too small for other detection means. Depending on the logic of the supporting electronic circuitry, either features which are darker than the surrounding background, or vice versa, can be detected. There may be several small features very close together, e.g., a 0.127-mm ladder chart.

Receiving light when dark is expected could indicate the incorrect orientation of a part before it reaches an automatic assembly station or an undrilled hole. The failure to receive light could indicate the absence of a minature component to the scanning unit.

Scanning units, connected to tachometers, retrieve information from confined spaces. Connected to counters, they operate in a similar way.

With the support of electronic devices, *miniature targets* can be scanned at *very high speeds.* Tachometers can operate up to 30,000 rev/min with TTL devices but up to 1,000,000 rev/min with C-MOS ones.

Not all the applications of fibre-optic scanning units are high-speed ones. In medicine, scanning units, with a lens system, detect small, relatively fast movements — muscle response, ocular pulse, blood vessel dilation, cardiac apex impulse and vibrations. The latter also has applications in engineering.

Scanning units are not restricted to opaque targets; they can be used on translucent ones. The two techniques are reflection comparison for opaque targets and light barrier for translucent ones.

The *higher thermal limits* of fibre optics enable them to have applications where electronic components cannot be used. Optical fibres can normally be used between -40 and 125°C. Special fibres can withstand 300°C (600°F). Many electronic components are confined to the -20 to 75°C range. When within that range receivers do not respond consistently.

Movements *towards* a probe or *across* it can be detected.

Data processing is another form of scanning.

3.6.2 Construction

All scanning units incorporate an optical-fibre probe, a light emitter and a light receiver. The probe can have segregated or mixed fibres. Figure 2.21 showed a circular probe with segregated fibres; Figure 3.11 shows three other arrangements with approximately identical cross-sectional areas. The dots represent receiving fibres in a mixed-fibre arrangement; they do not emit light. In all four examples the fibres are either segregated or mixed coaxial.

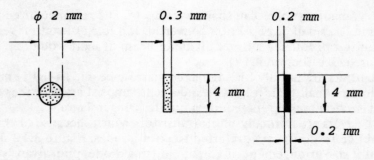

Figure 3.11 Probe end sections

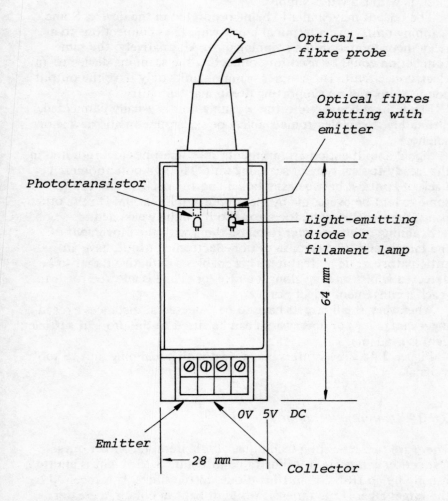

Figure 3.12 Layout of scanning unit (full size)

Rectangular probes can span two targets. The rectangular end-section is a small cross-section converter. If a longer length is required, an emitter producing a larger diameter beam of light would be required (see Section 4.12).

Emitters are usually high-radiance, narrow-beam, IR-light-emitting diodes, or small visible-light incandescent lamps. Larger probes or cross-section converters require a quartz halogen lamp.

Receivers are normally phototransistors which, because of their amplified response, are preferred to photodiodes. Figure 3.12 shows the full size arrangement of a light-emitting diode, phototransistor and the abutment ends of the probe.

The connections vary for these scanning units. Some are supplied ready for connection to a mains supply. They incorporate transformers and rectifiers. Other devices are supplied for building into circuits with a 5V d.c. supply.

The output may similarly be incorporated in the device. Some scanning units have a terminal block which has connections to a relay incorporated in the scanning unit. Alternatively, the same connection could be used for connecting the scanning device to an electronic circuit. The simpler scanning units only have the output connections and no supporting preliminary circuitry.

From these connections the scanning unit is usually connected, ultimately, to an electromechanical or electropneumatic device or solenoid.

Signal amplification arrangements are sometimes incorporated in the 'ready-to-use' type of scanning unit. This is a potentiometer. Lack of contrast between the target and the background can to some extent be overcome by electronic amplification. On the other hand, the signal may be too strong so it can be weakened.

Scanning units, in order to phase them with the movement of the targets and the response of non-electronic output, have an anticipatory or delay feature. This enables a defective target to be detected whilst moving along a conveyor. Time is allowed for it to reach a convenient reject point.

When very small targets have to be detected at distances exceeding 4 mm, a lens or lens system can be fitted to the probe if sufficient light is available.

Figure 2.44 shows filters being used with a scanning unit to split a signal.

3.6.3   Operating techniques

There are two operating techniques; both are non-contact ones. *Reflection comparison* is illustrated in Figure 3.13. Light is emitted, usually by an IR-light-emitting diode, to the fibres. It is received by the target area. If the target is white or light in colour, a greater amount of light is reflected to the receiving fibres, and ultimately to the abutting phototransistor, than is the case if the target is dark in colour. The logic of the scanning unit is that it differentiates between the desired and the undesired conditions by detecting the difference in the amount of light received. A missing component means that less light is reflected; a missing hole means that more light is reflected.

The *light barrier technique* is shown, enlarged, in Figure 3.14. Two optical-fibre probes are opposed. The emitter gives a beam of light. A larger cross-section converter would provide a curtain of light.

When a target obstructs the beam, or curtain, of light, the electronic circuitry responds according to whether it is in the desired or undesired

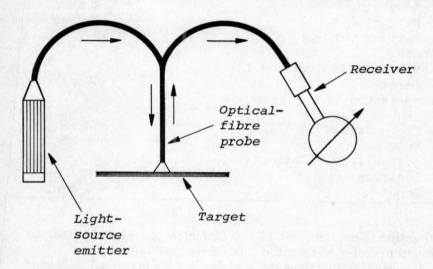

Figure 3.13   Reflection comparison technique

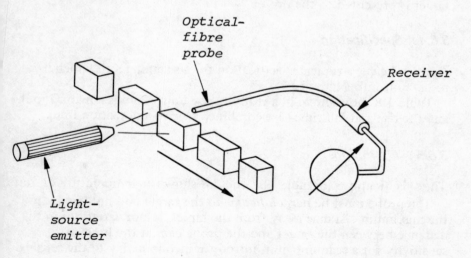

Figure 3.14   Light barrier technique

TABLE 3.7

Component	Value
Emitter life:	
Lamp, visible	40,000 hr
LED, infra-red	>40,000 hr
Peak spectral emission, LED	900 nm
Peak spectral emission, lamp	600 nm
Phototransistor:	
Rise time	14 μsec
Fall time	10 μsec
Maximum voltage	20 V d.c.
Probe:	
Sections as Figure 3.13	
Length	500 mm
Temperature:	
Probe, flexible	-40 to 125°C
Probe, rigid	-80 to 300°C
Body	0 to 75°C
Repeatability of target:	
Position	$\pm 10^{-7}$ mm
Input	5 V d.c.

condition. Since darkness, as well as light, is sensed, the technique is sometimes referred to as the dark barrier technique.

In certain applications the scanning devices may need strobing. This is effected by switching on the emitter immediately before a target is presented to the probe.

3.6.4 Specification

Table 3.7 gives a typical specification for a scanning unit which has to be built into a circuit.

Table 3.8 shows how, in a ready-to-use unit, there is a mains input and the rise and fall times are combined to give a detection time.

3.6.5 Limitations

The specifications in Tables 3.7 and 3.8 show the temperature limitatio

The probe must be *perpendicular* to the target and not less than the maximum distance away from the target. Minor variations in the distance between the target and the probe can, at the limit of sensitivity for a scanning unit, introduce inconsistency to the sensing. A target could be present, but undetected, if the probe were marginally too far away.

The probe is additionally subject to the *fibre limitations* described on page 36ff.

The shape of some components is such that one component can be *detected more than once,* e.g. springs. If the shape is regular, this can be overcome by calibrating on the basis of the number of pulses received. However, amorphous shapes need more than one probe to detect a combination of features. Some of the larger components can be detected by radar without fibre optics.

Atmospheric dust in the environment may contaminate the end of a scanning unit probe. Infra-red light reduces this because it is more energetic and penetrative. Low-power air jets and vacuum bleeds can overcome the problem.

Scanning units are limited in that few of them quantify information collected by their probes. Process control units which quantify their light input, by conversion, are described in the next section.

TABLE 3.8

Dimensions	63 x 100 x 95 x 110 mm
Weight	500 g
Voltage	24/48 or 110/240 V a.c.
Frequency	50/60 Hz
Power disipation	3.5 V
Emitter	Light-emitting diode
Oscillator frequency	approx. 6 kHz
Shortest detection time:	
Electric output	1 msec
Relay	8 msec
Switching frequency:	
Electric output	500 Hz
Relay	15 Hz
Life:	
Electronic	Indefinite
Relay	5×10^6 movements
Maximum switching power:	
Electric	+0.5 to 12 V
Relay	150 VA
Sensing contrast	Dark on light
	Light on dark
Target distance (without) lens	4 mm
Target size, rectangular probe	0.025 mm
Peak sensitivity	900 nm
Delays:	
Rise time	0.05 to 5 sec
Fall time	0.05 to 5 sec
Ambient temperature	-20 to 75°C

3.7.1 Applications

As their name suggests, these units are used to control processes. Like scanning units, they use the light barrier and reflection comparison techniques. They quantify the light received by using more sophisticated circuitry than scanning units. Additionally, they sense larger areas than scanning units. Targets are still slender, but longer ones can be sensed. Many process control units are used in conjunction with fast-moving conveyors.

In the food industry they are used, for example, to detect and initiate the rejection of burnt cornflakes which are amorphous shapes and have various shades of 'burntness'. In the same industry, a slow-moving application — unacceptable colour variations in the manufacture of instant coffee — are detected and corrected.

Continuous components, such as metal strips, are examined for local defects.

Car engine pistons are examined for roughness of surface finish as they leave the centreless grinding process. Crankshaft bearings are similarly checked at the place of manufacture on the machine itself.

One process control unit can supervise several stations. Coin blanks are checked, for surface irregularities, before being struck, at the rate of 23,000 pieces per hour by six detection heads.

Steel bars are dimensionally checked, and other engineering components, classified ones, are automatically gauged.

3.7.2 Construction

Figure 3.15 shows the general arrangement of a process control unit. The *power supply* has to be *stabilised* so as to provide a consistent optical datum. The stabilised light source, which can accept a 13 mm diameter lightguide, can illuminate a fibre area of 133 mm^2, i.e. a 0.5 mm continuous slit 266 mm long in a cross-section converter. The slit, however, might be allocated to several cross-section converters. Fibre optics is used to stabilise the light source. Light is bled, by a lightguide, from the main bundle of fibres to a photocell. This arrangement enables intensity differences of $\pm$ 0.02 per cent to be achieved.

Either mixed with the *emitting* fibres, or alongside them, are receiving fibres for *collecting and conducting* light from the target area to the photo receivers. Since visible light has to be used to achieve the requisite intensity and stability, silicon semiconducting material is used.

A modicum of visual control over the process is given by an *analogue indication dial*. However, the principal use of this is for setting up. The human eye does not respond sufficiently quickly to target information when the process is a fast moving one.

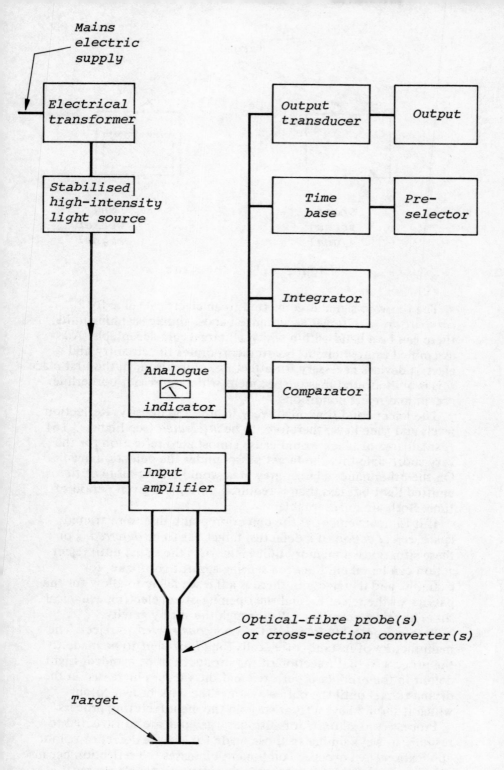

Figure 3.15 Process control unit

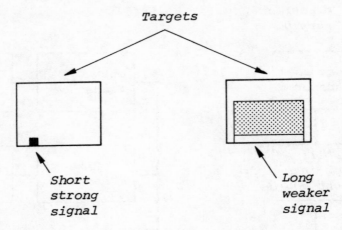

Figure 3.16 Integration

The received signal is converted to an electrical value for *comparison* with target acceptance bands. Unlike scanning units, there can be a band within which all targets are acceptable. An assembled printed circuit board incorporates the circuitry and electric devices necessary for effecting comparison. In the first place this is built around master specimens which represent borderline accept and reject conditions.

The reject condition might vary in area and density. Reflection levels and time have, therefore, to be *integrated* (see Figure 3.16). A small area of black would cause almost zero reflection for the very short time that the target passes under the detection head. On the other hand, a larger grey area would reflect more of the emitted light but less than is required during the given period of time. Both are unacceptable.

If it is inconvenient at the detection point, due to restricted space, *reject action* of a defective target has to be *deferred*. For these situations, a memory unit memorises the signal until reject action can be taken. Between sensing a burnt cornflake, for example, and its rejection, there is a 0.8 sec delay to allow for the passage of the reject to, and the opening of, an electromechanical slit. The reject cornflake falls through the slit by gravity.

Process control units are not always *constructed* to reject. The manufacture of instant coffee calls for *adjustment* to be made to the process so that rejection of the product can be avoided. Light colour in the product is detected and the vacuum increased in the drying tunnel until the coffee assumes the dark brown colour which it should have at that stage in the manufacturing process.

Process control units for automatic gauging are constructed to respond in ways similar to those made for surface defect or colour applications. An oversize component increases the reflection because it is nearer to the detection head; an undersize one decreases it.

The movement is towards or away from the detection head rather than across it.

3.7.3   Limitations

The limitations of process control are primarily mechanical. *Displacement* in the presentation of the target to the probe can introduce variations in the reflection received by a sensitive process control unit. A positional tolerance of ± 0.05 mm is required. The angle between the cross-section converter or probe and the target must be maintained. To achieve this, copper strip has to pass through additional idler rollers. For wider materials, such as metal sheet, detection heads have to be mounted on porous-pad air bearings. The detection heads, which are often cross-section converters, and less frequently probes, are fixed in one position wherever possible.

The throughput speeds of process control units are retarded by the speed at which the *mechanical handling* part of an installation can present and remove targets. Given faster handling, if that were possible, quantities in excess of 23,000 coin blanks per hour could be processed.

The applications of some process control units are described in more detail in the following chapter.

Four

FIBRE OPTICS FOR...

4.1 Résumé

The purpose of this chapter is to assemble and relate material from
the preceding three chapters to some, but not all, of the problems
that fibre optics can overcome or ameliorate. The situations are
intended to be representative of those to which fibre-optics technology
is applied. For that reason, where appropriate, the sub-heading
'Illustrative of' appears at the end of a sub-section. Under this heading
general uses of fibre optics are briefly stated so that the classification
of these applications, by industries and professions, will not introduce
a restrictive parochialism to the descriptions.

Wherever possible the 'aim' of an application is stated so that the
goal to which the fibre optics user is aiming is clearly identified. The
'equipment' to do this is next stated. The 'method' in which the
equipment is used is stated next.

Some of the descriptions do not always fit the above sub-headings.
Where the descriptions do not fit, other sub-headings have been
chosen.

4.2 Aircraft manufacture and maintenance

4.2.1 External inspection of a component

1 Aim To inspect components with a relatively broad surface for
surface cracks, without a microscope, but with magnification and
viewing angles which cannot be achieved with a bench magnifier (see
Figure 4.1).

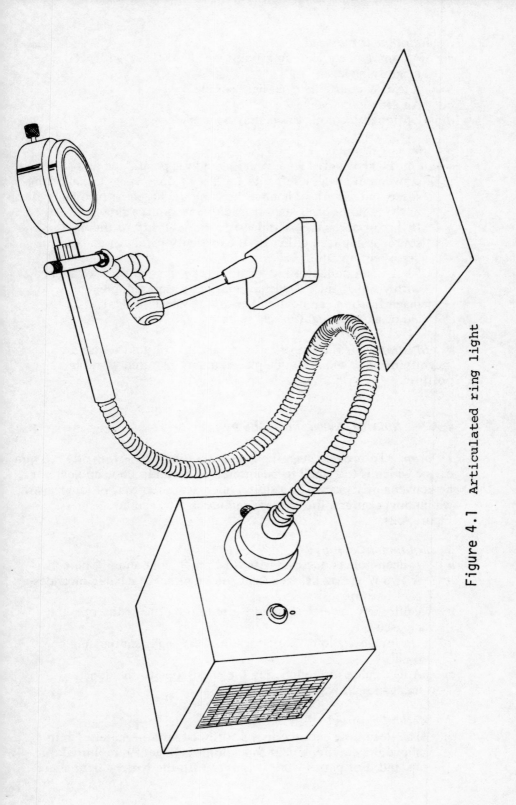

Figure 4.1 Articulated ring light

2   *Equipment required*
a   A complete circle or four-point ring light fitted with ✕10
    magnifying lens.
b   A 150-W quartz halogen light source.
c   An articulated stand.
d   Optionally, colour filters may be used.

3   *Method adopted*
a   On the articulated stand the ring light is positioned at a
    convenient angle, relative to the size of the inspection, allowing
    space underneath to handle the component, e.g. a turbine blade,
    in the region of the highest isophotes from the ring light.
b   If the surface is highly reflective, the intensity of the light
    source is adjusted either by the intensity regulator and/or the
    use of colour filters.
c   As the component is moved under the ring light it is examined,
    with visible light, for surface cracks. If required, greater
    magnification can be used because the level of light, with the
    aid of ambient light, is adequate.

4   *Illustrative of* the use of intense light, distributed round a
magnifying lens, which can be placed in any required working
position.

4.2.2   Internal inspection of 0.8 mm cavity

1   *Aim*   To provide magnified, well illuminated viewing of a 0.8 mm
cavity which is too small to be entered by an endoscope and cannot
be conveniently placed on a stereo microscope, so that residual swarf,
which might enter a fuel system and block jets, can be seen (see
Figure 4.2).

2   *Equipment required*
a   A deep-hole endoscope with ✕10 monocular and 50° objective.
b   A 150-W quartz halogen light source or an attachable incandes-
    cent battery.
c   Optionally, a vertical stand for mounting the endoscope may
    be used.
d   Optionally, colour filters for the 150-W light source may be
    used.
e   A lightguide for connecting the endoscope to the quartz
    halogen light source.

3   *Method adopted*
a   The deep-hole endoscope is positioned approximately 1 mm
    above the component. If the volume is large it is mounted on
    a stand. For patrol work it is used with the battery light source.

b   The image is a hollow inverted cone. If a through-hole is being examined there is a central dark area in the image. The information on the condition of walls, be they through or blind, is around the periphery of the image. Blind holes also have information in the centre, provided they are not so deep, e.g. 100 mm, that the bottom cannot be illuminated. With the battery light source, holes up to 25 mm deep can be examined.

c   If the light level is too high with the quartz halogen light source, it is reduced.

4   *Illustrative of* the ability of a wide-angle lens in a forward-viewing endoscope to collect an image of a laterally presented object from a cavity which is too small to enter.

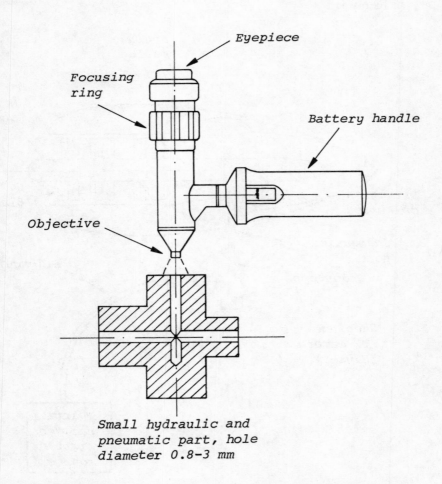

Figure 4.2   Deep-hole endoscope application

1 Aim To inspect turbine blades and combustion chambers whilst assembled in an engine, after running, for surface damage sufficient to warrant the engine being dismantled for their replacement (see Figure 4.3).

2 Equipment required
a For in-line access: a 7 X 300 mm endoscope with 30° objective and lateral-view X 10 magnification. Endoscopes incorporating a prism are often preferred. For curved access: a 8 X 1000 mm fibrescope, preferably with distal end, and an integral lightguide.
b A 150-W quartz halogen light source with either mains or petrol-driven generator supply.
c Optionally, a closed-circuit television camera and monitor may be used.

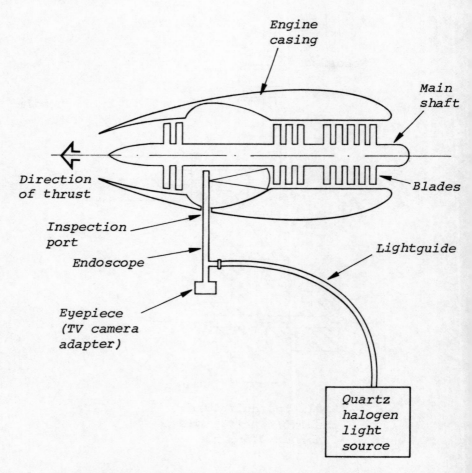

Figure 4.3 Internal inspection of jet engine

d   A lightguide for connecting the endoscope to the light source.

3   Method adopted   This application often requires more than one
man to rotate the engine.
a   The endoscope or fibrescope is eased through the inspection
        part and the inspection area located.
b   The quartz halogen light source is set for maximum illumination.
c   If a television system is being used it is focused onto the object.
d   If blades are being inspected, the engine is slowly turned so that
        each blade can be seen either through the monocular or on the
        television monitor.
e   If a fibrescope is used the end is moved to ensure that the whole
        of the object area is viewed.

4   Illustrative of the ability of an endoscope to collect information
on an object some distance away, e.g. 300 mm from the objective, in
an environment where there are carbon deposits.

4.2.4   Internal inspection of sheet metal assemblies

1 . Aim   To illuminate the inside of curved sheet metal assemblies,
e.g. fan components for 1:1 viewing of the welds in a situation where
a fibrescope is not justified.

2   Equipment required
a   A lightguide of 10 synthetic fibres 1 mm in diameter in a loose
        PVC jacket, 2 m long.
b   Light sources : *(i)* a 150-W quartz halogen with additional
        infra-red filter or *(ii)* a 20-W quartz halogen light source.

3   Method adopted   The lightguide is eased into the hollows of
the sheet metal assembly, and then moved up and down and
rotated so that potentially defective areas are illuminated.

4   Illustrative of the use of a lightguide to take intense light round
sharp corners and possibly over sharp edges, to assist the naked eye.

4.2.5   Other applications

1   The use of a synthetic monofibre, eased down an electro-
chemically drilled hole, to see whether short holes at right-angles
break through. If they do break through there is a speck of light.

2   The use of a ring light, with a magnifying lens, for free hand-
inspection of relatively small areas on a large component, e.g.
a turbine disc.

3   Inspection of aerial interiors is possible with a small diameter endoscope.

4   Inspection of curved core holes in aluminium castings is possible with a fibrescope (see Section 4.15.2).

5   Blind landing system instruments often use optical fibres.

4.3   Automotive vehicle manufacture and maintenance

4.3.1   Instrument illumination

1   Aim   To illuminate all dashboard instruments from a single lamp (see Figure 4.4) and reduce space.

2   Equipment required   A length of optical-fibre ribbon with glow areas, screen-printed legends (white on black) and white plastic backing of suitable width and length. The length should not be longer than necessary. One lamp for illuminating the ribbon.

3   Method adopted
a   The glow areas are formed by a hot stamping process which damages the fibre cladding so that light escapes. A heated die under a load of 4.5 kg (10 lb) pressing against an abrasive surface, e.g. smooth emery paper, is used. Dwell times and die temperatures have to be established empirically.
b   A 'sandwich' is made of the legend, ribbon and plastic backing. Letters on the legend should be at least 2 mm high.
c   The 'sandwich' is threaded through the spaces and fixed under the appropriate instruments.
d   At the lamp end an eyelet is fitted to the 'sandwich' for crimping to the lamp.

4   Illustrative of the use of fibre optics to eliminate several lamps and save space for the designer with low-intensity light.

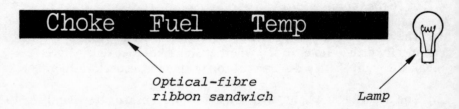

Figure 4.4 Dashboard illumination

1 Aim To provide illumination and magnified viewing of the seat, sack and spray holes so that defects which could be environmentally detrimental, or adversely affect engine performance, can be identified.

2 Equipment required
a An endoscope, 4.6 × 60 mm, with the appropriate objective, (i) seat 40°; (ii) sack 10°/30°; (iii) spray holes 10°; ×10 magnification. With a suitable endoscope, a lateral revolvable mirror tube can be fitted for viewing the breakthrough of inlet holes.
b A 150-W quartz halogen light source.
c A lightguide for connecting the endoscope to the light source.
d Optionally, a silicon diode camera tube and television monitor for high magnification, e.g. ×100, and easier presentation of the image may be used (see Figure 4.5).
e An endoscope vertical-mounting stand.
f Optionally, colour filters for bright machined finishes may be needed.

3 Method adopted
a The nozzle is offered to the endoscope which is fitted with a collar to prevent damage to the fibre annulus and objective. If television is not used, the eyepiece is uppermost, whereas if television is used the objective is uppermost.
b For nozzles in the black condition, the light source is at maximum

Figure 4.5 Inspection of diesel nozzle injection with television. The inner circles are drill marks and the four spray holes are in the centre

intensity. Otherwise the light intensity may have to be reduced.

c   In the eyepiece the image is a solid circle: *(i)* the seat appears as a chamfered annulus with the more distant sack and spray holes, in the centre, possibly out of focus; *(ii)* the sack image has a tapering appearance; *(iii)* the spray holes have a solid, shallow dished image. The endoscope may need focusing separately for the seat, sack and spray holes. The image on the television monitor may be somewhat flatter than it is in an eyepiece, which has greater depth of field.

4   *Illustrative of*
a   The ability of endoscopes, with different objectives, to collect different information from the same component.
b   The alternative positions of endoscopes to facilitate the handling of a small component.

4.3.3   Brake hose internal inspection

1   Aim   To provide magnified viewing of the internal structure of 3-mm bore hose on a sample basis.

2   Equipment required
a   A 3 X 185 mm endoscope, forward-viewing, 30° objective magnification X 10. Wider-angled objectives are not available for small diameter endoscopes. The maximum length is 185 mm.
b   A 150-W quartz halogen light source.
c   Alternatively, a deep-hole endoscope, if the hose is on bottom limit, may replace the endoscope in *(a)* above.
d   A lightguide for connecting the endoscope to the quartz halogen light source.

3   Method adopted
a   The 3-mm endoscope is gently eased into the hose since it is very slender.
b   The image is a long slender cone with the information at the periphery and a central dark spot where all the available light is absorbed. The deep-hole endoscope produces a similar but shorter image.
c   Lubricants used in the manufacturing process improve the reflectivity of the black object area.

4   Illustrative of the ability of a forward viewing endoscope to collect 'sliding fit' information from laterally presented objects.

4.3.4   Other applications

1   Monitoring of head, side and rear lamp malfunctioning, using

2   Process control of surface finish of petrol engine pistons.

3   Inspection of cylinder interiors using an endoscope or fibrescope after removal of a spark plug.

4.4   Carpet manufacture

4.4.1   Pattern reading

1   Aim   To read 3-mm squares printed in two colours on a continuous strip of paper so that the pattern can be reproduced, in larger format, by a loom, (see Figure 4.6).

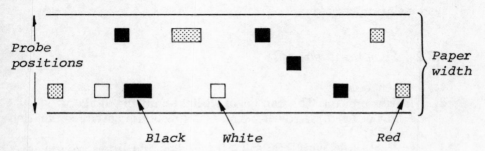

Figure 4.6   Carpet pattern reading

2   Equipment required   A bank of scanning units each with one infra-red emitter but the receiving probe split into two branches, one having a colour filter.

3   Method adopted
a   The probes are fixed over the pattern square positions and the emitter and receivers built into a circuit.
b   The probes read black — which absorbs the light, red, which partially absorbs it, and white which reflects it.
c   The three positions are connected to needle positions on the loom.

4   Illustrative of the use of non-contact opto-electronic scanning units to improve the ease of making patterns and manufacturing speeds in a traditional industry.

4.5.1   Scope

Fibre-optic telecommunications embraces both the audio and visual
forms, over relatively short distances, using optical-fibre link-ups.
With the aid of repeaters, messages have been transmitted up to 8 km.

On land, telephones, even videophones, inter-office links, between
and within buildings, and information retrieved from regionally
situated data banks may all, within the foreseeable future, be using
fibre-optic telecommunications.

A colour television programme has been passed through a 1.25 km
land line before being nationally broadcast.

In the air, intercoms, using fibre optics, are used for communication
in some military aircraft. At sea, development work is being undertaken into the use of fibre optics for the on-board communications
systems of supertankers.

4.5.2   Equipment required

1   An encoder.
2   A light emitter. This can be an IR-light-emitting diode or
    a shorter lived, narrower beam, faster responding semiconducting
    laser.
3   Optical fibres with 0.02 to 0.10 mm diameter fibres assembled
    into bundles and further assembled into cables with a possible
    polystyrene strengthening member. These multi-mode fibres
    will probably supersede the single-mode fibres which have both
    handling and preparation difficulties.
4   A receiver: a pin or avalanche photodiode.
5   A decoder.
The words encoder and decoder are general terms used to describe
the pieces of equipment which are the first and last stages in the
conversion of the audio-visual input to, and from, the infra-red light
which actually travels through the optical fibres (see Figure 4.7).

4.5.3   Method adopted

1   The input is converted by the encoder to electrical signals
    which represent either the sound waves of the voice or the
    scanning of visible media.
2   The emitter sends out probes of infra-red light corresponding
    to the electrical values, in strength and duration.
3   The infra-red light is launched into the fibres which conduct it
    to the receiver.
4   The receiver re-converts the light to electrical values.

5   The decoder completes the re-conversion process by returning
      the output to the input form.

4.5.4   Advantages

a   Fibre-optic communications systems have a large bandwidth,
      e.g. 1 GHz. The bandwidth is the maximum rate at which
      information can be transmitted.
b   Immunity to electromagnetic interference in electrically noisy
      situations.
c   High security against 'tapping'.
d   Much greater flexibility than the majority of waveguides.
e   Low weight when compared with copper — 60 per cent less.
f   Ability to resist vibration.
g   Glass fibres have no fire risk.
h   Inability to form unwanted earth loops.
i   Inability to short-circuit adjacent filaments when fractured.
j   High capacity — one laser beam can carry up to 100,000
      telephone conversations. Thirteen polystyrene optical fibres
      have replaced 300 copper wires.

4.5.5   Limitations

a   Coupling losses can cause substantial attenuation. Dead space
      at the emitter/fibre and fibre/receiver junctions and (unless
      optically corrected) the beam spread of 7° associated with
      semiconducting lasers are the usual sources of launching
      problems. Mono-mode fibres are particularly prone to launching
      losses because it is difficult to produce an accurate square end.
b   Jointing and cabling, in order to produce longer lengths, are
      currently receiving development attention.

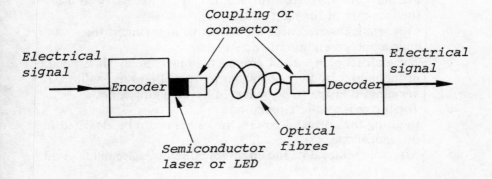

Figure 4.7   Fibre-optic communications

c   Although attenuation in fibres has been dramatically reduced, it can have a cumulative effect. The sound loss is currently 2 dB/km (3.2 dB/mile) while the laser pulse spread is 0.3 nsec in 50 m.

Since fibre-optic telecommunications, or communications, are attracting a large research effort, this section has been restricted to a précis form.

4.6   Construction, architecture and building

4.6.1   Cavity inspection

1   Aim   To examine cavities for bridging and deterioration of the insulation, for defective welds in box sections but without damaging the structure.

2   Euipment required
a   A large diameter endoscope, e.g. 14.5 mm, with wide-angle lens (×5 magnification) and, usually, lateral view. The length of the endoscope depends on the distance of the object from the access hole.
b   A 150-W quartz halogen light source.
c   A twin-armed lightguide. One arm is connected to the endoscope and the other arm is free. Each arm is 1.5 m long.
d   Optionally, a television camera and monitor may be used to improve image presentation.
e   Optionally, a photographic camera, for recording purposes, may be used.

3   Method
a   Two or, if necessary, more holes are drilled in the wall. The endoscope is eased through one hole and the object viewed. If the light level is too low due to absorption by a rough masonry or rusted steel surface, it may be necessary to use the free arm of the lightguide.
b   This arm is inserted near the object to supplement the endoscope's own illumination system.
c   The endoscope is moved around the object area. The free arm of the lightguide may be moved to other holes to enable a meaningful image to be collected by the endoscope.
d   The image is a solid circular one.
e   By using the correct adapters the cameras can be attached to the endoscope.
f   After the removal of the endoscope the holes are made good.

4   Illustrative of
a   The steps which can be taken to counter inadequate illumination

when using a wide-angled endoscope where the object areas are of low reflectivity.

b   The internal examination of structures without damage.

4.6.2   Development visualisation

1   Aim   To produce, from an architect's model, a large enough image to enable the effects of proposed developments to be assessed by interested parties.

2   Equipment required

a   A modelscope with extra-wide-angle 70° lens which simulates the objective of the human eye. The modelscope usually, but not always, has optical fibres because there is sometimes difficulty in getting ambient light between the model and the objective.

b   A 20-V or 150-W quartz halogen light source, if the modelscope has fibres.

c   A lightguide for connecting the modelscope to the light source.

d   A television camera and monitor, often a large one.

e   Optionally, a stand for holding the modelscope and camera may be used.

f   Optionally a photographic camera may be used to provide a permanent record.

3   Method

a   The modelscope and camera are positioned outside or on the 'building' which has to be visualised.

b   Interested parties — conservationists, town planners, traffic engineers, investors, architects and members of the public have the image presented, in large format, on the monitor in a three-dimensional form.

4   Illustrative of the use of a fibre-optic viewing instrument to collect an enlarged image from a small object area, to which there is easy access, and to relay it simultaneously to several people, with the aid of a complementary instrument.

4.7   Data processing

4.7.1   Input illumination

1   Aim   To illuminate, by one lamp, a card, so that holes can be detected by small opposed photodiodes when they receive light through holes in a punched card.

2   *Equipment required*
a   A cross-section converter (see Figure 2.19) with suitable
      spacing and slit area and a single-input integral lightguide.
b   One lamp.

3   *Method adopted*
a   The input end of the cross-section converter, which is similar
      to a circular lightguide, is connected to the lamp.
b   The other end, the cross-section converter, is fixed 2 - 3 mm
      above the card reading position.

4   *Illustrative of* the, light from one lamp being distributed to
several close, in line illumination points on miniature targets. Very
high intensity is used without adversely affecting the performance
of the electronic components.

4.7.2   Input illumination and retrieval

1   *Aim*   To illuminate and detect codes in punched cards and tape
using the light barrier technique.

2   *Equipment required*
a   An illumination cross-section converter (see Figure 2.19). The
      slits match the receiving cross-section converter in number,
      size and spacing.
b   A multi-branch, multi-slit receiving cross-section converter
      (see Figure 2.17).
c   One lamp — preferably stabilised.
d   Photodiode receivers corresponding to the number of branches
      of the receiving cross-section converter.

3   *Method adopted*
a   The two cross-section converters are positioned opposite to
      each other, on either side of the card or tape.
b   The illumination cross-section converter is connected to the
      lamp.
c   The multi-branch receiving cross-section converter branches
      are connected to the photodiodes.
d   When there is a hole, the light shines through it into the relevant
      slit in the receiving cross-section converter, then to the photo-
      diode. The electrical resistance is reduced and the signal
      interpreted by the supporting electronic circuitry.

4   *Illustrative of* the use of a bank of opposing emitting and
receiving slits in a restricted space connected to a lamp and photodiodes in an unrestricted space.

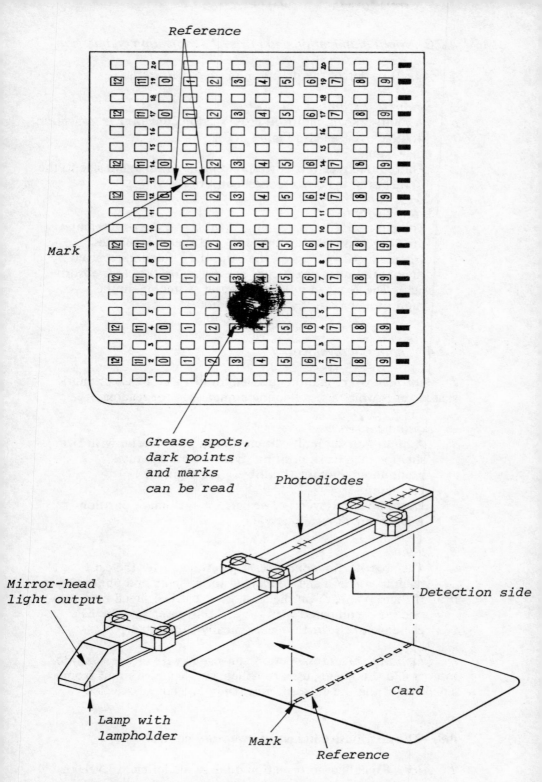

Figure 4.8   Improvement of data processing input

1   Aim   To detect the marks on mark-sensed cards.

2   Equipment required
a   A multi-branch, multi-slit cross-section converter as in Section 4.7.2 above (see also Figure 2.17).
b   One lamp.
c   Photodiode receivers corresponding in number and size to the branches of the cross-section converter.

3   Method adopted
a   The cross-section converter is fixed above the card reading position, connected to the lamp and the photodiodes.
b   When a mark appears on the card, the reflection received by the photodiodes is reduced. The marks on the cards absorb light. The changed reflection causes switching, in the supporting circuit, by the photodiodes.

4.7.4   Input improvement

1   Aim   To overcome the problem of dirt and grease on mark-sensed cards which use reflection comparison for reading.

2   Equipment required
a   A multi-branch, multi-slit cross-section converter with two slits for each mark position. Each slit incorporates randomised illumination fibres (see Figure 2.17)
b   One lamp.
c   Photodiode receivers — two sets for each mark position.
d   A large mirror (see Figure 4.8).

3   Method
a   The cross-section converter is positioned over the card reading position and connected to the lamp and photodiodes.
b   If there is a mark on the card, the electrical signal from the one slit is compared with the adjacent reference slit. The presence of the dirt is electronically eliminated.

4   Illustrative of a cross-section converter (with close-proximity interrupted slits) being used to collect two kinds of input from a small target and the adjacent area. Only one lamp is needed.

4.8   The distributive industry: inventory control

1   Aim   To collect information quickly on deficient levels of

consumable foods at the point of sale, without writing, which
introduces possible transcription errors, and is slow.

2   *Equipment required*   A fibre-optic pen (see Figure 4.9). The end
section is circular with segregated optical fibres (see Figure 2.21).
The pen contains an emitter which may be a visible lamp or IR-
light-emitting diode if infra-red printing inks are used for the codes.
The receiving fibres are connected to a castor-mounted memory

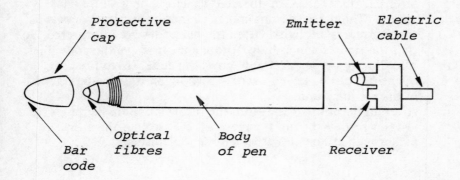

Figure 4.9 Fibre-optic pen (the two parts
are separated to show emitter and receiver)

unit. The electricity supply is from a battery. The following is a
typical fibre-optic pen specification:

Scan speed   0 to 760 mm/sec with 5 lines/mm
Orientation   The read range, and geometry of the tip,
                   allow operation at tilt angles from 0 to
                   $45°$ at a distance of 0.8 mm

The resolution versus distance for 1.8 line pairs/mm is:

| Distance, | Tilt | |
mm	$0°$	$45°$
1.27	88–93	89–94
6.35	91–98	86–90
12.70	91–98	–
19.00	97–94	–

The protective tip and the angle affect the distance.

3   *Method adopted*
a   A bar code (see Figure 4.10) is fixed on each shelf. The codes
      could equally be on books, mail, detergent cartons or
      pharmaceutical containers. The code comprises black lines of

Figure 4.10   Bar code

different thicknesses, at different spacings, on a white background. The permutations and combinations of thickness and spacing represent digits. Often the bar codes are interpreted. So that the figures which they denote appear above the code lines.

b The fibre-optic pen is swept across the code. If required, quantities can be keyed in for recording on the magnetic tape in the memory unit.

c After an iteration of the store the information on depleted stock levels is retrieved from the memory unit so that the stock pickers can arrange for stocks to be replenished.

4.9   The electrical industry

4.9.1   Instrument assembly

1   Aim   To present a low-magnification image of the components in an instrument, e.g. helical gear chains, as an assembly aid, while allowing adequate space for the assembler to make adjustments to the instrument.

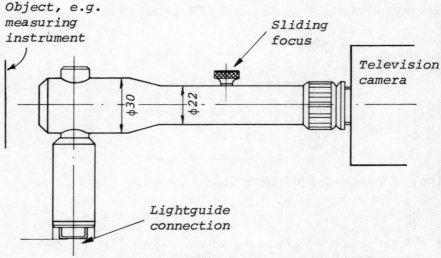

Figure 4.11   Television probe

a   A television probe incorporating C-mount and a focusing system
    (see Figure 4.11).
b   A 150-W quartz halogen light source.
c   A lightguide for connecting the television probe to the light
    source.
d   A television system using a simple vidicon tube and small
    monitor.

3   *Method adopted*
a   The probe is positioned 100 to 150 mm away from the target
    components in the instrument facing away from the assembler.
    The television monitor faces the assembler.
b   The focusing slide is moved until an acceptable image appears
    on the monitor.
c   The assembler then works inside the instrument whilst watching
    the monitor.

4   *Illustrative of*
a   The use of fibre optics as an assembly aid, in conjunction with
    closed-circuit television.
b   The ability of fibre optics to provide a smaller-sized lens system
    than is possible with a television camera.

4.9.2   Instrument malfunction

1   *Aim*   To provide a magnified view of closely assembled
components, e.g. in a meter, but without taking either heat or
electricity into the object area, so that dismantling for inspection and
subsequent re-testing can be avoided.

2   *Equipment required*
a   An endoscope 4.6 × 125 mm with 45° objective, × 10 magnifica-
    tion and preferably mirror tube for connection to lateral view.
b   A 150-W quartz halogen light source.
c   A lightguide for connecting the endoscope to the light source.

3   *Method adopted*
a   The endoscope is eased inside the instrument near components
    which are potentially troublesome. For lateral viewing it is
    rotated.
b   Foreign bodies and damaged components are seen in the solid
    circular image conveyed by the eyepiece.

4   *Illustrative of* a small diameter endoscope being used, on medium
distance objects, for industrial diagnostic purposes.

a   Scanning units are used for detecting the presence of filaments for automatic assembly into electric lamps.

b   Synthetic fibre lightguides are used to illuminate the inaccessible parts of electrical and electronic control cubicles.

4.10   Electronic component manufacture, positioning and assembly

4.10.1 Assembly of electronic devices

1 Aim   To detect incorrect orientation of, for example, a capacitor washer, which is matt grey on one side and shiny black on the reverse, and to correct it before reaching an automatic assembly station.

2 Equipment required

a   A bowl feeder with exit channels of suitable width for the washers.

b   A scanning unit with a circular mixed-fibre probe, relay output, infra-red emitter and signal amplification.

c   An automatic assembly station.

d   An air jet (see Figure 4.12).

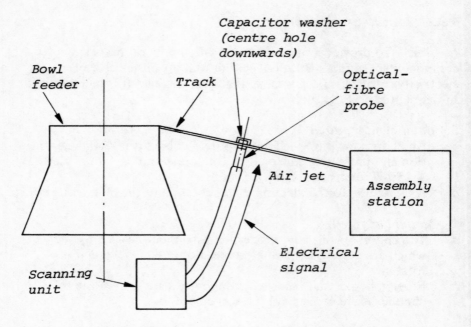

Figure 4.12 Scanning of capacitor washers

3   Method adopted

a   The washers leave the bowl feeder face downwards.

b   The scanning unit is strobed by a micro-switch to scan when a washer is over the probe.

c   If matt grey is detected the electrical resistance in the photo-transistor is higher than when shiny black is detected.

d   When matt grey, the unwanted condition, is detected by the scanning unit an air-operated solenoid blasts a small jet of air onto the washer. The air blast turns the washer about its axis whilst still in the feed channel. Another jet of air across the channel halts the other washers while the incorrectly orientated one is turned.

4   Illustrative of the use of a scanning unit using reflection comparison to orientate a component where the distinguishing features are only marginally different. Both sides are dark but one is matt, the other shiny and thus more reflective.

4.10.2   Printed circuit board assembly

1   Aim   To prevent the misplacing of components on printed circuit boards in medium-volume situations.

2   Equipment required

a   Synthetic mono-fibres of several colours, 1-mm diameter approximately 500 mm long.

b   Component hoppers corresponding in number to the devices which have to be placed on the board.

c   A mounting frame for mono-fibres and printed boards. The frame incorporates a template with holes identical in diameter and position to those in the board.

d   A 20-W quartz halogen light source.

e   A single 3 X 700 mm swan neck (see Figure 4.13). (Three millimetres is the active diameter of the fibres, not the outside diameter of the swan neck.)

3   Method adopted

a   The mono-fibres are placed in the template holes and slightly heated at the ends to anchor them.

b   The other ends are clipped, with cable clips, to the hoppers containing the components which are to be placed in the corresponding positions on the board. Colours are used to indicate polarity. It is unnecessary to have every hole for an integrated circuit illuminated.

c   When the template and hoppers have been linked, the unassembled board is placed over the template.

d   The swan neck is positioned over the first hopper so that the

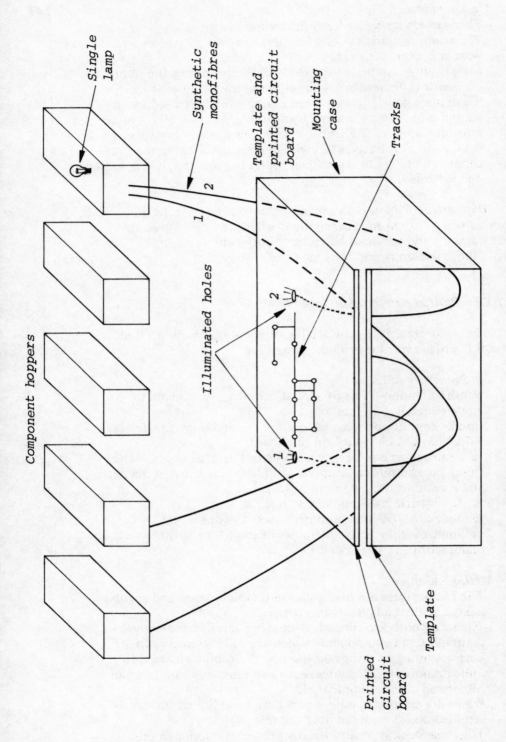

Figure 4.13 Printed circuit board assembly

fibres receive light which they transmit to the hole positions in the template. The light shines through these into the holes on the printed circuit board above.

4   *Illustrative of* the use of optical fibres with a low-power light source to provide a simple assembly aid for an operation where incorrect assembly could remain undetected until final test.

4.10.3   Printed circuit board positioning

1   Aim   To illuminate very confined spaces so as to enable loose connections and potential short circuits, e.g. solder drips, to be seen.

2   *Equipment required*
a   A lightguide comprising five or ten 1 mm diameter, 2 m long, synthetic mono-fibres, optionally in a PVC jacket.
b   A 150-W quartz halogen light source to illuminate the lightguide.
c   An additional infra-red filter for the light source.

3   *Method*
a   The mono-fibres are threaded into the narrow gap between the underside of the board and the mating assembly (see Figure 4.13).
b   Where additional illumination is required some of the cladding is scraped from the fibres, near their ends.
c   The illumination from the synthetic fibres is sufficiently intense for the positioning of the board to be viewed from the edge and, where boards are translucent, from above. In the latter case features are silhouetted against the board.

4   *Illustrative of*
a   The use of synthetic fibres in spaces too small for other fibre-optic viewing instruments.
b   The use of intense quartz halogen light to shine through materials.

4.10.4   Other applications

1   Scanning units, with suitable probes, and sometimes with lenses, are used for the automatic assembly of devices which are manufactured in large quantities, e.g. transistors.
2   Continuous-slit cross-section converters connected to a 150-W quartz halogen light source are used for the edge illumination of printed circuit boards to reveal broken tracks. The light penetrates approximately 70 mm into a translucent board.
3   Deep-hole endoscopes are used for the magnified inspection of

individual through-plated holes. Sometimes an illuminated box is used underneath.

4   Ring lights, with X10 magnifiers are used for less thorough inspection of through-plated holes. Several holes are seen at one viewing. But there is little depth of focus.

4.11   The food and drink industry

4.11.1   Burnt cornflake detection and rejection

1   Aim   To detect and automatically reject burnt cornflakes. They vary in size, shape and amount of damage. They are presented on a 1 m wide white conveyor moving at between 2.5 and 3 m/sec.

2   Equipment required
a   A process control unit with a high-intensity stabilised quartz halogen light.
b   Enough continuous-slit cross-section converters, connected to the process control stabilised light source and the photoreceivers, to span the conveyor.
c   A memory unit for the reject signal.
d   A cornflake spreading bar in front of the cross-section convertors. so as to ensure that only one layer is presented to the cross-section converters.
e   An electronic integrator.
f   Electromechanical reject slits (see Figure 4.14). refers.
(The above equipment is similar to that for other process control

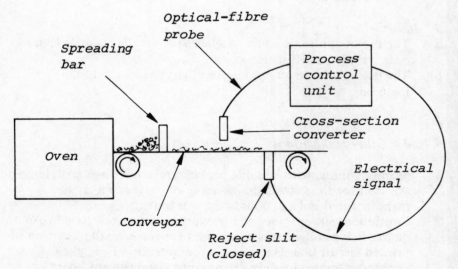

Figure 4.14   Burnt cornflake rejection

3   Method adopted

a   The cross-section converters are fixed over the conveyor, at a
        point where it does not move up and down.

b   Using master specimens, top and bottom limits for a band of
        acceptability are programmed into the process control unit.

c   When burnt specimens pass under the cross-section converters,
        the amount of reflected light is reduced; it is absorbed by the
        darker colours of the burnt flakes.

d   The reject signal is stored in the memory unit for about 0.8 sec.
        The time varies with the speed of the conveyor. This delay
        allows time for the defective cornflake to reach a position over
        an electromechanical reject slit and for the slit to be open in
        readiness for its arrival. The defective flake falls into the open
        slit. From time to time some good cornflakes enter the electo-
        mechanical slit when they are adjacent to burnt ones.

4   Illustrative of the use of fibre optics for a fast-moving fragmentised
product with three sets of variables — size, shape and discolouration.

Peas, dried haricot beans, potato crisps and chipped potatoes all
have spot defects which could be identified and rejected by the same
installation. Fragmentised non-ferrous metal scrap can also be sorted
by colour after cleaning in a fluidised bed.

4.11.2   Faulty lid detection

1   Aim   To detect insufficient or excessive deposits of sealing
compound in lid channels, before curing, and to stop the machine
when necessary. Lids rotate under the compound-dispensing guns
at linear speeds of approximately 4.5 m/sec at throughput rates of up
to 180 lids per minute. The compound colours are off-white or a
light brown.

2   Equipment required

a   A scanning unit with a rectangular probe, delayed rise time and
        IR-light emitter — and solid-state electronic output.

b   A compound-dispensing gun, lid, rotating vacuum collet and lid
        conveyor (see Figure 4.15).

3   Method adopted

a   The scanning unit probe is fixed 4 mm above the lid channel
        immediately behind the compound-dispensing gun. Lids are
        moved to the dispensing position by a rising cam.

b   When the lid is in the dispensing position, at a given signal, the

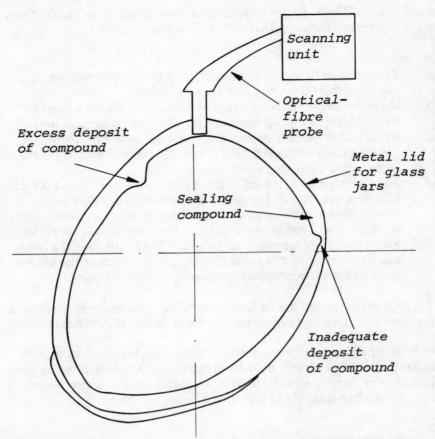

Figure 4.15 Scanning of sealing compound levels

needle valve in the dispensing gun is raised and the scanning unit switched on.

c   If there is insufficient compound the level of reflection is lowered. On the other hand, if there is an excess, as when there is overlapping at the end of the compound-dispensing cycle, excessive reflection is received by the fibres.

d   The two signals are interpreted by supporting electronic circuits. Both are undesired conditions.

e   The circuitry has an output to a magnet; this raises a leaf spring on the conveyor leading to the curing oven. The leaf spring tips the machine until the faulty lid is removed by the machine minder.

4   *Illustrative of* the use of a scanning unit to distinguish between two kinds of information collected by an optical-fibre probe from a miniature target.

1   Aim   To compare changes in the colour of brown instant coffee with pre-set standards, throughout the manufacturing cycle, and to initiate corrective action when the standards are not being achieved.

2   Equipment required
a   A process control unit with thermally protected probe.
b   A heated coffee manufacturing vacuum tunnel (see Figure 4.16).

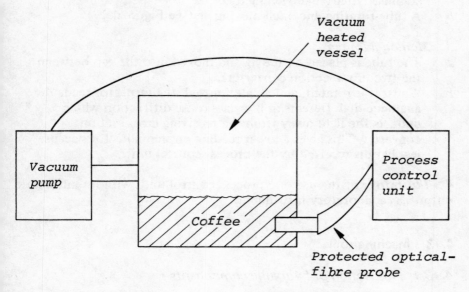

Figure 4.16   Instant coffee manufacture

3   Method adopted
a   The optical fibre probe is fixed in the vacuum tunnel just below the level of the coffee.
b   When the coffee is lighter in colour than the pre-set levels for the various check points in the manufacturing cycle, the process control unit gives an output signal to the vacuum pump. It responds by increasing the vacuum in the tunnel. Increasing the vacuum darkens the colour of the coffee. Each pulse increases or decreases the vacuum by 0.1 torr (1 torr = 0.0193368 lbf/in^2).

4   Illustrative of the use of a process control unit to avoid rejecting faulty material in a slow-moving process. (This particular process usually takes about 7 hr.)

4.12.1 Fluorescent lamp glass tubes

1 Aim To produce an analogue indication of the presence of air lines, hair-lines, bubbles and inclusions in sample glass tubes. These tubes are used in fluorescent lamps.

2 Equipment required
a   A process control unit with an analogue indicator, an opposing positioned continuous-slit cross-section converters and a stabilised quartz halogen light source.
b   A tube-rotating and axial-moving jig (see Figure 4.17).

3 Method adopted
a   The tube is placed on the jig and moved into the gap between the two cross-section converters.
b   Whilst it is rotated, and axially moved, the inspector reads the analogue dial. Defects in the glass cause diffraction which deflects the light away from the receiving cross-section converter. This gives a lower reading on the indicator because less light is received by the process control unit.

4 Illustrative of the use of a process control unit, without automatic output, as a laboratory instrument.

4.13 Machine tools

4.13.1 Inspection of hydraulic components

1 Aim To inspect confined spaces and small cavities for the presence of swarf or burrs. These fragments of metal could enter hydraulic oil and impede the operation of machine control circuits.

2 Equipment required There are a number of alternatives, depending on the ratio of the cavity diameter to length (see page 59):
a   Magnified viewing: *(i)* a hole that *can* be entered — an endoscope of suitable length and diameter with 45° objective, × 5 or × 10 magnification and optionally with mirror tube for conversion from lateral viewing to forward viewing is used. (If the access is curved, a fibrescope may be required as the endoscope will be unable to enter the cavity.) *(ii)* a hole that *cannot* be entered by a normal endoscope. A deep-hole endoscope with 50° objective can 'see' into holes less than 3 mm in diameter if there is in-line access (see page 94).
b   For viewing, *without* magnification, in a hole that can be entered,

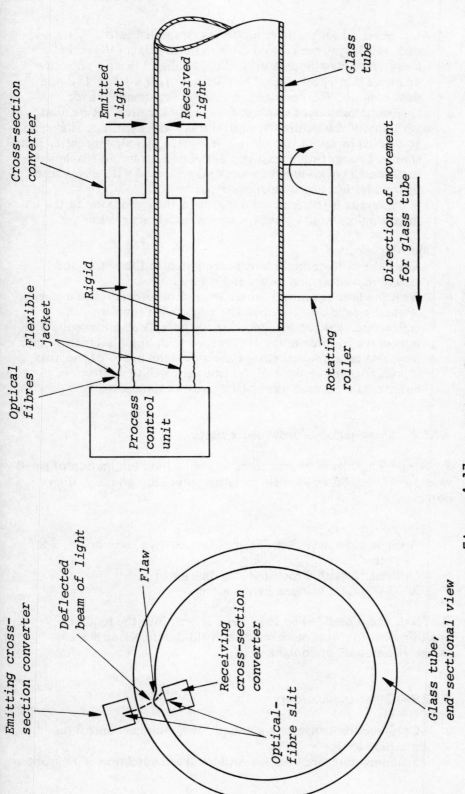

Figure 4.17 Checking of glass tubes

use either *(i)* synthetic mono-fibres assembled into a lightguide of 5 or 10 arms each 1 mm diameter either in a PVC jacket or loose, *(ii)* a glass-fibre, PVC-jacketed lightguide with an active diameter 2 × 1800 mm with or without *(iii)* a 1.5 × 150 mm rigid conduit. The conduit can have an attachable mirror.

c   Quartz halogen light sources: *(i)* A 150-W source can be used with any of the equipment but, if used with synthetic fibres, an additional infra-red filter is required with this high-intensity light source. Colour filters may also be needed for bright machined surfaces. *(ii)* A 20-W light source may be used with any of the lightguides but not with endoscopes.

d   A lightguide is required for connecting the endoscope to the quartz halogen light source if magnified viewing is the aim.

3   Method adopted

a   The lightguide or endoscope is presented to the object and slowly moved up and down and rotated.

b   With the hole-entering endoscopes and fibrescopes, the image is usually solid and circular. The image with the deep-hole endoscope, which does not enter the hole, is a hollow cone if it is used for through holes. If it is used with blind holes the image may also be a solid tapering cone unless the depth of the hole is greater than the limit of the endoscopes illumination. In the latter case the image has a black circle in the centre.

4.13.2   Illumination of small workpieces

1   Aim   To provide intense, precisely directed illumination of small workpieces, e.g. on watchmakers' lathes, precision grinders or jig borers.

2   Equipment required

a   A single swan neck 3 × 350 mm arm, or two swan necks, 3 × 750 mm.

b   Optionally, with a focusing lens on each arm.

c   A 20-W quartz halogen light source.

3   Method adopted   The swan neck is bent into the required position over the workpiece so as to avoid dazzle, shadow and excessive amounts of coolant.

4.13.3   Other applications

a   Cross-section converters are used for numerical control on machine tools.

b   Scanning units are used for drill-breakage detection. The probe is positioned at the point of the drill which is strobed when it is in the retracted position.

c   Scanning units are additionally used on transfer lines for detecting undrilled through holes prior to tapping. This avoids the consequences of attempting to tap an undrilled hole.

4.14   Medicine

4.14.1   Endoscopy

In the medical context, 'endoscopy' includes both rigid and flexible fibre-optic viewing instruments. On occasions, it also includes non-fibre-optic instruments such as sigmoidoscopes.

1   Instruments available
a   Magnified viewing instruments:
   i   Basic instruments: bronchioscopes, oesophagogastroduo-denoscopes, gastroscopes, colonoscopes, laryngoscopes and vaginascopes.
   ii   Variations: forward or lateral viewing, objective angle, lengths (colonoscopes vary between 1000 and 1900 mm).
   iii   Ancillary features: fibrescopes may have a moveable distal end, forceps, dilators, needles, electrocautery tips and tubes for gas insufflation and extraction, liquid introduction and retrieval and anaesthetic introduction.
b   Light sources: a 150-W quartz halogen light source with a fail-safe feature (see page 89).
c   Complementary instruments: principally television and photographic cameras. Progress is being made with the use of colour television cameras attached to fibre-optic viewing instruments.

2   Uses
a   Diagnostic:
   i   Medical fibrescopes are complementary to radiology and the longer established rigid endoscopic instruments. They provide an alternative diagnostic procedure when the more usual ones are negative or inconclusive.
   ii   They permit the visual examination of body channels, without surgery.
   iii   Biopsies and brush and smear specimens can be taken under visual supervision.
   iv   Gases can be extracted from the body channels or insufflated, e.g. oxygen, or carbon dioxide — whichever is appropriate.
   v   Dyes and other liquids can be introduced.
b   Therapeutic:
   i   Cauterisation.

      ii   Prosthesis.
      iii   Removal of foreign bodies.
      iv   Application of clips.
      v   Introduction of drugs.

c   *Post-operative:*
      i   Inspection of the repaired area.
      ii   Removal of sutures.

3   *Advantages*
a   Direct visual examination of the affected areas can be made.
b   Patient comfort and convenience is ameliorated.
c   Anaesthesia is not required although sedation or analgesia is advisable.
d   Specimens can be taken under visual supervision.
     Visual aids can be prepared for training students.

4   *Limitations*
a   Endoscopists require extensive training.
b   Endoscopy cannot replace radiology because control of the object end of the instrument is imperfect.
c   Where there is inflammation or suspected perforation it is inadvisable to use endoscopy. In some post-operative conditions the same limitation must be observed.

4.14.2 Movement detection

The application of a scanning unit to the detection of movement is described on page 110.

4.14.3 Blood analysis

Ultra-violet light is transmitted through quartz fibre cross-section converters in some blood-analysis machines. Some oscillographic recorders, which are used for echocardiography, incorporate fused optical-fibre faceplates.

4.15 Metal manufacture

4.15.1 Blanks

1   *Aim*   To detect surface defects on coin blanks of various shapes, materials, diameters and thicknesses at a rate exceeding 20,000 pieces/hr with ±1 per cent repeatability.

2 Equipment required

a   One process control unit with six optical-fibre detection probes, a stabilised light source, electronic integrator and a memory unit. In front of each probe is a lens to make the system more insensitive to lateral movements.

b   Material handling equipment for taking the blanks to the sensing stations, and for rotating and turning them under the detection heads (see Figure 4.18).

3 Method adopted

a   The process control unit is programmed with master blanks to identify the top and bottom limits of the acceptance band. Separate printed circuit boards are assembled for the different combinations of variables.

b   The blanks are fed to the six sensing positions. There they are rotated four times before being turned. They are again rotated four times after turning.

c   The integrator 'averages' minor changes in the blank presentation position and assesses short deep scratches and long thin ones for acceptability.

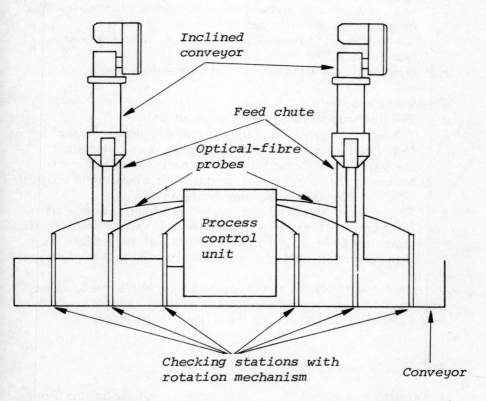

Figure 4.18   Automatic checking of coin blanks

d If the level of reflection is below the required level for a blank to be accepted, a reject signal is given to the memory unit.

e The signal, after allowing time for the passage of the reject blank down the exit chute, moves over a gate to deflect the reject blank into the reject bin.

4 Illustrative of the use of one fibre-optic process control unit with *six* sets of detectors or probes on a fast-moving automatic process. Each station checks one blank per second. Unlike the cross-section converters over the cornflake conveyor these six probes do not span the blank. They collect information from a small area.

4.15.2 Castings

1 Aim To inspect curved core holes, which will become oil ways, in castings, for the presence of sand or flash which could, if allowed to remain, enter an aircraft fuel system.

2 Equipment required

a A 6 × 870 mm forward-viewing fibrescope with 40° objective and a resolution of 30 line pairs/mm, enclosed in a woven braid or spirally wound metal jacket.

b A 150-W quartz halogen light source with variable intensity (see Figure 4.19).

c Optionally, colour filters may be used with the light source.

3 Method adopted

a The fibrescope is eased into the curved cavities.

b The light intensity of that light source may need reducing. Newly cast surfaces, being rough, can produce reflection in many directions. These reflections irritate the human eye. Moreover, the colour filters can introduce a measure of contrast which enables sand to be more clearly seen.

c The image in the fibrescope eyepiece is a long, slender, curved, tapering one. It may be solid or hollow. If the curves are gentle there may be irregularly shaped central dark spots in the image. If the curves are light, the image may be solid with no dark spot.

4 Illustrative of the use of a forward-viewing fibrescope to collect an image of a laterally presented object to which there is a curved access and sharp edges between the access and the object.

4.15.3 Hollow forgings

1 Aim To assess whether internal cracks in a large hollow forging are small enough to permit machining to begin, e.g. a gun barrel.

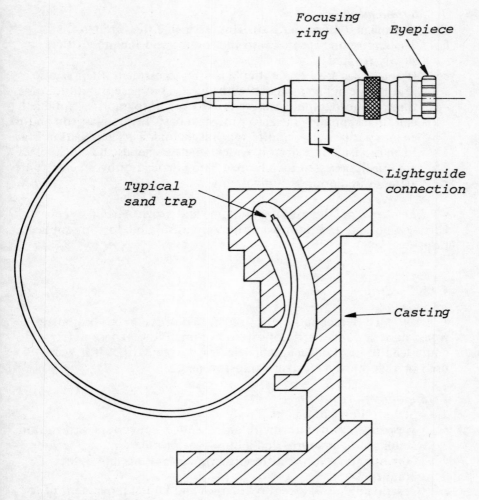

Figure 4.19 Fibrescope in curved-casting core hole

Focusing ring

Eyepiece

Lightguide connection

Typical sand trap

Casting

2   *Equipment required*
a   Magnetic powder.
b   A large diameter endoscope of suitable length, e.g. 14.5 × 2000
     mm with a 45° objective , lateral viewing, × 10 magnification
     and a graticule eyepiece.
'c   Optionally, a right-angled adapter for turning the monocular
     eyepiece through 90° may be fitted to the endoscope.
d   A jig for easing the endoscope into the forging.
e   A lightguide 1.5 m long for connecting the light source to the
     endoscope.
f   A 150-W quartz halogen light source (see Figure 4.20).

3   *Method adopted*
   a   The magnetic powder is distributed inside the forging.
   b   The endoscope is eased into the forging and the mirror tube
         slowly rotated.
   c   If cracks are seen the graticule is used to estimate their size so
         that they can be compared with the level of acceptability. Some-
         times a photographic camera is used to photograph acceptable
         and unacceptable cracks to provide a basis for comparison and to
         ensure parity between different inspectors. This application is a
         form of visual integration. Crack lengths, shapes and thicknesses
         are jointly assessed by a human being instead of by an electronic
         element in a process control unit.

4   *Illustrative of* the use of a graticule and a right-angled eyepiece.
Ultra-violet light and fluorescent inks are unsuitable for this application.

4.15.4   Strip

1   Aim   To detect and indicate surface defects on continuously
rolled narrow copper strip of different dimensions. These defects
could lead to local unacceptable electrical irregularities if it were
used in high-tension electrical transformers.

2   *Equipment required*
a   Strip-rolling equipment.
b   A process control unit with cross-section converters, integration,
      stabilised light source and a library of printed circuit boards
      corresponding to the different strip dimensions and defect
      conditions.
c   A strip guide to prevent oscillation and lateral movement near
      the cross-section converters.
d   A strip-cleaning device.

3   *Method adopted*
a   The process control unit is programmed with an acceptability
      band of defect depths and areas.
b   The cross-section converters are positioned on either side of the
      strip after the strip-cleaning device as it emerges from the
      rolling machine.
c   When a defect of either unacceptable depth or length or area
      appears, an electropneumatic cylinder edge marks the edge of
      the strip adjacent to the defect after receiving a signal from the
      process control unit to initiate the marking.

4   *Illustrative of* the use of a fibre-optic process control unit, working
by reflection comparison, simultaneously over two broad surfaces of

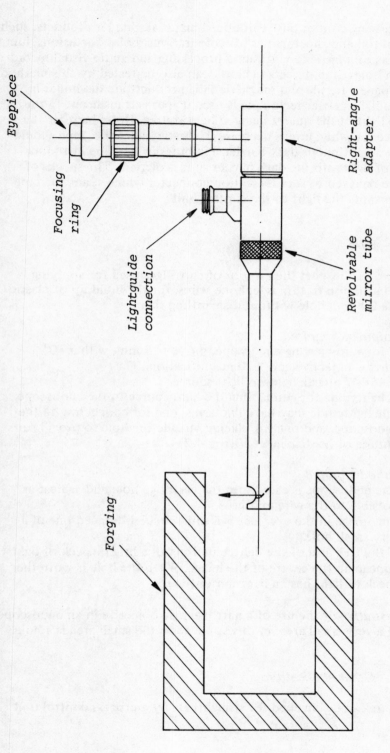

Figure 4.20 Hollow-forging inspection

a fast-moving continuously produced material. Similar products, such as metal foil and sheet, paper (discolourations, holes, inclusions, lumps and creases), plastic tape for data processing and audio-visual recording (pin holes) and defects in cloth, can also be tested by this method.

The paper, textile and magnetic tape products are dissimilar in that the light barrier technique is used. Paper is translucent to the intense light of the quartz halogen light source. Holes increase the translucency while lumps decrease it. Plastic 'magnetic' tape should provide a continuous dark barrier. Pin holes in the tape introduce specks of light into the dark barrier so it is pierced. The specks of light are received by a cross-section connecter which spans the tape and transmits the light to the control unit.

4.15.5 Wire

1 Aim To inspect the interior of cartridges used for applying plastic insulation to thin telephone wires, for the build up of plastic near the 0.8-mm hole without dismantling them.

2 Equipment required
a A forward-viewing endoscope, 4.6 × 125 mm, with a 10° narrow objective and ×10 magnification.
b A 150-W quartz halogen light source.
c A lightguide for connecting the light source to the endoscope. The lightguide may have two arms, one for connecting to the endoscope, and one for placing outside the hole to provide an outline of its shape (see Figure 4.21).

3 Method adopted
a The endoscope is eased into the cartridge hole and as near as possible to the wire exit hole.
b The image in the eyepiece is a solid circular detailed one of a very small hole.
c If the free arm of the lightguide is used, a bright speck of light appears in the centre of the image. When the hole is worn the speck of light has an irregular outline.

4 Illustrative of the use of a narrow-angle objective in an endoscope to inspect a very small area; in this application the small area is a hole.

4.15.6 Other applications

Steel bars can be checked for straightness by a process control unit.

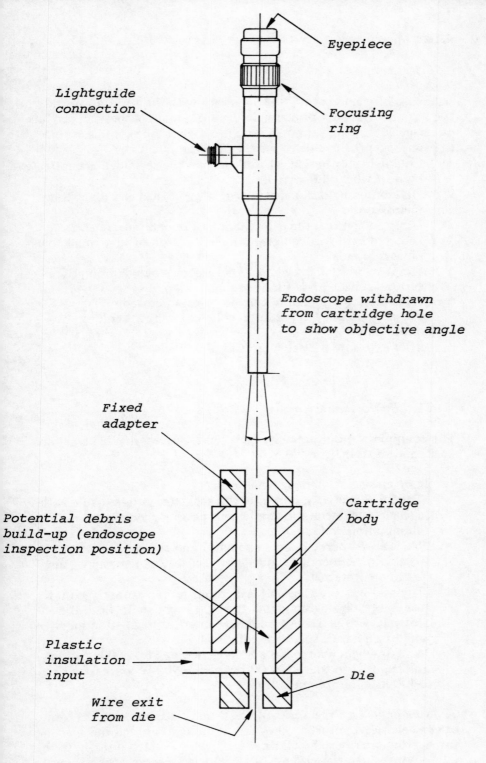

Figure 4.21   Wire-drawing cartridge

4.16.1 Aims

When used for microscopy the aims vary with the application and type of microscope. For many applications microscopes can produce an image without the aid of fibre optics. When fibre optics is used, the aim is to provide one or more of the following:

1   Cold light, for biological work so that the specimen is unaffected by heat when illuminated.
2   Distributed light, for metallurgical, geological and dissection microscopes.
3   Directed light for the examination of opaque specimens.
4   Directed light for high-power magnification of very small objects or areas.
5   Supplementary light when a microscope's own illumination system is inadequate for a specimen.
6   Contrasted illumination, for geological microscopy.
7   Constant colour temperature, coherent light for colour photography.
8   Brief exposure of specimens to intense light.

4.16.2 Equipment available

The equipment is in four categories: light sources, ring lights, swan necks and flexible lightguides.

1   Light sources   All quartz halogen:
a   The 20-W source has no intensity regulator. It is suitable with a single swan neck for supplementing a microscope's own illumination system.
b   The 150-W source is large enough for all the applications but its colour temperature and slightly shadowed light may cause *(c)* to be preferred.
c   The 500-W source has an ideal colour temperature of 3200 K for colour photography and the light is coherent. Both the 150 and 500-W light sources have intensity regulation facilities and often have a space for colour filters.
d   A 250-W light source has a homogenous 2,500,000 lx output (as opposed to the 1,500,000 lx of the 500-W source) and a 3400 K colour temperature.

2   Ring lights (see Figures 2.14, 2.15 and 2.30) are a special form of cross-section converter. They distribute light around the specimen for stereomicroscopy. Small ring lights, which attach directly to the microscope's high-power lens, use the aperture angle of the fibres to 'inject' light into small spaces between the specimen and the objective — a maximum distance of 4 mm.

3   *Swan necks* are solid fibre bundles. When a single swan neck is used, it supplements a microscope's own illumination system (see Figures 2.31 and 2.32). The twin-armed swan neck is used to illuminate a specimen, from either side, and to either eliminate or position shadows. It is the sole source of illumination. A three-armed swan neck has two arms illuminating the specimen and one illuminating the microscope lens system.

The deep hole endoscope can sometimes be used as an alternative to a microscope (see page 94).

4.17   The oil, gas and chemical industry

4.17.1   Corrosion and crack inspection of pipework

1   Aim   To identify corrosion and cracks in, for example, welding inside pipework. Access is often awkward when the pipework is installed in a plant.

2   Equipment required
a   A forward-viewing endoscope, e.g. 14.5 X 1000 mm, with X 10 magnification and 45° objective. The length is restricted by the need to manouevre on catwalks and ladders. The length, although less than that of the longest endoscopes, is usually adequate to view the reaction areas inside pipes where damage is most probable.
b   A safe power source for fitting directly onto the endoscope, e.g. a lead acid accumulator which can be carried on a belt fastened around the inspector's or surveyor's waist. The need for high-quality images usually precludes the use of fibrescopes.
c   Optionally, a photographic camera may be attached to the endoscope. Weight considerations preclude the use of all but the small, poorer-resolution, boiler-tube inspection television cameras. These television cameras do not use endoscopes.

3   Method adopted
a   The endoscope is eased into the tube, after the removal of the plug or header.
b   The image is long and slender. At the limit of the light source illumination there is a dark spot in the centre of the image.
c   Photographs are taken when a permanent record is required.

The applications described in this section are similar to those described elsewhere. The first two require a process control unit, the third an endoscope.

4.18.1   Outsorting of contraries from dry waste paper

The aim is to identify and reject non-paper substances (which cannot be outsorted by other means) from dry fragmentised waste paper. The equipment used is almost identical to that for the rejection of burnt cornflakes (see Section 4.11.1). The conveyor is replaced by a vacuum conveyor from which the fragmentised material is suspended as it moves past the cross-section converters. When a contrary is detected an air jet blows it downwards from the conveyor.

4.18.2   On-line inspection of finished paper during reeling, for inclusions, holes, lumps, etc.

The aim is to mark the paper where there are defects. The equipment used is similar to the process control unit for the continuous checking of copper strip (see Section 4.15.4) except that paper is translucent. If the cross-section converters come into contact with oscillating paper, they need a hard, smooth abrasion-resistant face, proud of the continuous slit in the cross-section converters to avoid damaging the polished fibres.

4.18.3   Drying-cylinder inspection in paper mills without dismantling the plant

An endoscope of suitable length and diameter is used.

4.19   Pharmaceuticals

4.19.1   High-speed tablet counting

1   Aim   To count tablets at the point where they leave the die at speeds of up to 500 per second.

2   Equipment required
a   A high-speed tablet press.
b   A scanning unit with circular probe, infra-red emitter and electronic output connections.
c   Optionally, nickel cadmium batteries may be incorporated

to provide a continuous electrical supply if tamper-proof counting is required.

3   Method adopted

a   The probe is fixed inside the machine near the point where the tablets are ejected from the die, e.g. 10 mm away.

b   The scanning unit is connected to an electronic counter calibrated to work on the basis of 1 pulse = 1 tablet.

c   Precautions are taken against dust impeding the sensitivity of the probe by having a low-pressure jet of air blowing across its end.

d   The nickel cadmium batteries are enclosed in the scanning unit's metal case so that its electrical supply, and that of the counter, cannot be interrupted. The case is fastened and the probe is sealed in position for the production of tablets that could have a 'Black Market' value.

4   Illustrative of

a   The use of an optical-fibre probe to capture meaningful information at the point of origin.

b   The means by which a scanning unit can be made tamper-proof.

4.19.2   Other applications

1   The fibre-optic pen page (see Section 4.8.1) could be used to read bar codes on pharmaceutical product lables.

2   Process control units could be used to detect irregularly shaped and unevenly coated tablets.

4.20   Photography

This section describes how fibre optics assists photography, and not vice versa. In several places reference has been made to the use of photographic cameras with fibre-optic viewing instruments. Three ways in which fibre optics assist photography are described in this section.

4.20.1   Subject illumination

1   *Aim*   To provide continuous intense illumination so that shadows can be identified and light intensity measured, before photographing a small subject.

2   *Equipment available*   The area to be illuminated affects the choice:

a   Illumination devices:

    i   Ring lights — full circle of four point (see Figures 2.14 and 2.15) — for subjects of approximately 50 mm diameter and for holding round the camera lens.

    ii   Twin swan necks for subjects which may not be circular, e.g. trapezoidal. Focusing lenses concentrate the light on a 10-mm diameter circle. The two circles can be merged to illuminate non-circular subjects. A single swan neck can be used for a small subject where the elimination of shadow is not a requirement.

b   Light sources: these are all quartz halogen. They vary according to the illumination device and whether colour photography is being undertaken.

    i   A 20-W light source is adequate for a single swan neck even if a focusing lens is used.

    ii   A 150-W light source has the capacity for all the illumination devices.

    iii A 250 or 500-W coherent-light source is required for colour photography and the provision of shadow-free light at the attachment face of the light sources.

All three large light sources have intensity regulation. Many of these larger sources have a place for colour filters which can be used to introduce contrast into the subject.

c   A stand for holding and positioning the ring light or twin swan neck near the subject.

d   A light meter.

3 Method adopted   These are common to photography. The illumination is arranged to highlight particular spots, to eliminate or cast shadows and to suppress dazzle. Measurements are taken in order to calculate shutter speeds and distances.

4.20.2 'Frosting' for contrast

1 Aim   To emphasise a subject, by introducing contrast into the negative.

2 Equipment required

a   An optical-fibre fish tail (see Figure 4.22).

b   An incandescent lamp.

3 Method adopted   The areas on the negative, around the subject, are lightly touched with the fish tail, before fixing. This introduces tiny specks of light. The fish tail can be fixed to the lamp with an epoxy resin.

1 Aim   To remove any irregularities before fixing the negative.

2 Equipment required
a   Synthetic 1-mm thick mono-fibre.
b   An incandescent lamp.

3 Method adopted   With great care, the mono-fibre is touched to those parts of the negative where irregularities appear.

4.21 Security

4.21.1 Investigation of cavities

1 Aim   To examine irregularly shaped cavities (which are potentially hiding places for stolen goods and contraband) through a restricted access, without damaging the container in which the cavity appears or inconveniencing the general public. To examine cavities containing inflammable liquids, e.g. vehicle fuel tanks.

2 Equipment required
a   A battery light source for screwing directly onto a fibre-optic viewing instrument.
b   A fibre-optic viewing instrument of medium diameter and medium length.
  i   In-line access: a forward-viewing endoscope, 7 X 350 mm, with 30° objective and X5 magnification. The light source is insufficient for a larger objective angle than 30° while the field of view of a 10° objective is too small.
  ii   Curved access: a jacketed swan neck 9.5 X 215 mm fibre-scope or uniscope suitable for pre-forming before easing into the cavity.

3 Method adopted
a   The fibre-optic viewing instrument is eased into the cavity

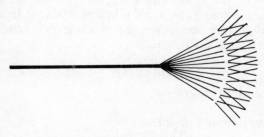

Figure 4.22 Fish tail

through a suitable access.

b   The instrument is moved backwards and forwards and rotated.

c   If the image in the eyepiece is suspect, the cavity is further investigated — not necessarily with a fibre-optic viewing instrument. Avoidance of inconvenience to the owner of the goods in the container may no longer be an aim!

4   Illustrative of the use of portable fibre-optic viewing instruments in 'one-off' situations.

4.21.2   Continuous unobtrusive surveillance through restricted access

1   Aim   To use a narrow restricted access for the insertion of a fibre-optic viewing instrument so that events in, e.g. a small room, can be presented to several specialists simultaneously — police, customs officials, behavioural scientists.

2   Equipment required
a   Fibre-optic viewing instrument:
   i   In-line access to a cavity: a forward-viewing endoscope of suitable length and diameter, with a 45° objective.
   ii   Curved access to a cavity: a fibrescope of suitable length and diameter, preferably with a distal end to direct it to the different parts of the cavity.
b   A low-light television camera and monitor. (NO light source is used because the light emitted by the fibres would attract attention.)

3   Method adopted
a   One person eases the fibre-optic viewing instrument through the access and focuses it while another, from in front of the television monitor, issues instructions as to the positioning of the instrument.
b   From time to time the fibre-optic instrument may be moved so as to survey another part of the cavity.

4   Illustrative of the complementary use of fibre-optic viewing instruments and a television system in low ambient light conditions, where space does not permit the use of a television camera without a fibre-optic viewing instrument.

4.22   Shipbuilding and repair

4.22.1   Diesel engine inspection

1   Aim   To inspect cylinder liners, piston crowns and the area behind the injector for wear, without dismantling the engine.

2 *Equipment required*
a   An endoscope of suitable length and diameter with a 30°
    objective, ✕5 magnification and the following mirror tubes:
    (i) lateral 90° ; *(ii)* retro 70°; *(iii)* prograd 115°; or four
    endoscopes for the different viewing directions if prisms
    are used.
b   A 150-W or larger quartz halogen light source.
c   A lightguide 1.5 m long for connecting the light source to
    the endoscope. Larger engines may require the use of an
    intrascope.

3 *Method adopted*
a   The injector is removed and the endoscope eased through the
    hole.
b   The following viewing angles are required for the different
    objects: *(i)* piston centres − forward; *(ii)* piston edges − prograd;
    (iii) liners − lateral; *(iv)* behind injector − retro.
c   The mirror tubes are rotated for lateral, retro and prograd
    viewing so that the whole of the object area is viewed.

4 *Illustrative of* the use of a large diameter endoscope with four
different viewing angles.

4.22.2 Other applications

1   Instrument illumination:
    a   Ring light, for marine compass illumination (see Figures
        2.14 and 2.15).
    b   Fish tail, for less critical instruments (see Figure 4.22).
    c   Optical-fibre ribbon, for legends and small instruments
        (see Figure 4.2).
2   Pipework inspection with an endoscope (see Section 4.17.1).
3   Localised magnified inspection of external welds (see section
    4.2.1).with a ring light and magnifying lens.
4   Injection nozzle inspection (see Section 4.3.2) with an
    endoscope. Since the object is near the endoscope the same
    diameter instrument, but slightly longer, is suitable, i.e. 4.6
    ✕ 125 mm.

4.23 Sport

4.23.1 Large format, immediate visual result display

1 Aim To make available to spectators the names and results of
competitors, either contemporaneously with, or immediately after,

an event, in high ambient light conditions.

2 *Equipment required*

a   Circular bundles of glass optical fibres arranged in a raster (see Figure 4.23) so as to be capable of representing up to 23 alphanumeric characters.

b   A driver unit connected to the console and measuring equipment and to the lamps.

c   Several quartz halogen lamps connected to bundle ends. One lamp may illuminate more than one bundle end. The figures 4, 8 and 9 share the same raster positions, like calulator photodiode digi

d   A magnifying lens to fit over bundle ends on the notice.

e   Heat-absorbing filters are needed for large quartz halogen lamps.

f   Lamps for illuminating the raster.

3 *Method adopted*

a   Competitors' names are pre-set on the console, some of which have small monitors.

b   In sophisticated display systems, the timing and measuring system is integral with the notice. A competitor touching the edge of swimming bath or breaking a beam of light could 'freeze' the time on the notice with the lamps supplying those points on the raster lit as instructed by the driver unit.

c   In less sophisticated display systems, competitors' times have to be keyed into the console before the raster points are illuminated.

4   *Illustrative of* the use of fibre optics to communicate variable information, clearly and quickly, in alphanumeric form.

The same principle is used for traffic control signs on motorways. The information there is either symbolic, ϕ (all clear) or numeric

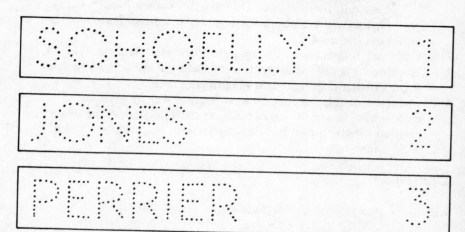

Figure 4.23   Raster (matrix) for sports results

(30). It remains illuminated for longer periods than information on sports events and there are fewer variables.

4.23.2   Other applications

1   Two cross-section converters are used to give the reference light points for ballistic tests on golf balls.
2   A battery-powered light source, with semi-rigid synthetic fibres, if used for the 'back lighting' of rifles and shot guns to search for dirt which collects in the barrel whilst out shooting.

4.24   Watch and clock manufacturing

4.24.1   Orientation of parts

1   Aim   To detect incorrect orientation of industrial fasteners, e.g. rivets, and to trip an automatic assembly machine.

2   Equipment required
a   A bowl feeder.
b   A scanning unit with a rectangular end-section probe, an instant signal and relay output.
c   An automatic assembly station (see Figure 4.24).

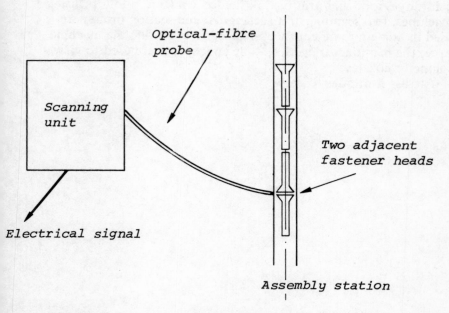

Figure 4.24   Incorrect orientation of parts

3   *Method adopted*

a   The probe is positioned so as to span the complete length of each fastener as it descends the chute from the bowl feeder.

b   If a fastener is descending, head first, a condition in which it cannot be assembled, two heads are detected by the probe. This condition provides an excessive amount of reflected light.

c   Reflection above the pre-set level closes the relay, which has been set to switch when 'light' is sensed, as opposed to 'dark'. Closing the relay trips the machine.

d   The machine minder then attends to the chute before re-starting the automatic assembly machine.

4   *Illustrative of* the use of a rectangular probe to span a thin miniature target.

4.23.2   Workpiece illumination

The equipment for illumination of workpieces is a swan neck and a 20-W quartz halogen light source (see Section 4.13.2).

The list of applications of fibre optics in this chapter may, at first sight, appear long. Yet nothing has been mentioned of the use of fibre optics with the colorimeter which is used to match paint, dye, food and printing ink shades. On large colour printing machines, two scanning unit rectangular end-section probes are used in tandem to detect sheet misfeeds from inside the machine. In textile manufacturing, deep-hole endoscopes are used to view spinneret nozzles . . . .

The list continues to grow.

CONCLUSIONS

There are indications that the use of fibre optics is not 'levelling out'. More areas of industrial, commercial, professional and domestic activity are increasingly using fibre optics.

Organisations with problems similar to those briefly described in the final chapter of the book, are adopting those (tried and tested) solutions. Growth in the use of fibre optics is being stimulated by the quest for economic growth. The persistent need to achieve improvements in quality and/or quantity, whilst at the same time using less, and certainly no more, resources than previously, offers fibre optics an opportunity to compete with other technologies.

If a process can be partly, or completely, automated, labour costs are reduced. Material is saved if scrap is eliminated or quickly identified. A machine, producing more, as the result of better control, enhances its utilisation. Fibre optics is already active in optimising the use of these three factors of production — manpower, material and machines — by visual and non-visual means.

Besides improving the use of material, fibre optics are replacing materials. Copper is being replaced in some telecommunications systems. Writing is being superseded in some inventory control procedures. Endoscopic examinations are a substitute for dismantling and some destructive tests.

Developments in materials — optical glass, transparent plastics, lamps and electronic devices, will push back those frontiers which currently limit, or impede, the application of fibre optics. Technological improvement is a continuing process. May it accelerate and improve the ability of fibre optics to respond to the needs of industrial, commercial, professional and domestic activity.

Appendix One

SOME FIBRE-OPTICS MANUFACTURERS

AER Optics
American Cytoscope Manufacturers
American Optical Company
Barr & Stroud
Bausch & Lomb
Bell Telephone Company
Corning Glass Works
Dolan-Jenner Industries
Du Pont
Dyonics
Eden Instrument Company
Electro Surgical Instruments
Elmhurst Instrument Company
Fish Schumann
Fort Electronique
Fuji Optical Company
Galileo Corporation
General Electric Company

Hellermann Deutsch
Hird-Brown
Incom
International Fibre Optics
ITT
Jenaer Glaswerk Schott
Machida
Nippon Sheet Glass Company
Olympus
Philips
Pilkington
Plessey
Rank Kershaw
Skan-A-Matic
Charles R. Thackray
Volpi
Welch Allyn
Richard Wolf

Appendix Two

OPTICAL GLASS

There are more than 300 different types of optical glass but there are two basic forms: *crown glass* (which, in some ways, is similar to window glass) with a refractive index between 1.45 and 1.6 and *flint glass* with a refractive index between 1.55 and 1.9. The cooling rate, after annealing, affects the refractive index.

The variations between glasses are in terms of density, hardness, discolouration and dispersion. To crown and flint must be added fluosilicate flint, hard crown, fluoborate, phosphate and germanate optical glasses. They have few principal constituents but many trace constituents to improve their optical and mechanical properties.

The constituents are heated in a platinum-lined tank, since molten glass attacks refractory materials. Flint glass is heated to 1400° C, crown to 1550° C. Much of the glass used for fibre optics is poured into moulds and continuously annealed. After cooling, the slab is broken for the selection of pieces free from inclusions and stresses.

Lens manufacture involves roughing operations, with a shaped tool for curved faces, on an oscillating head, after suitable diameters have been trepanned from the broken pieces. The lens is rotated throughout the operation. Abrasive powders are used for both roughing and finishing operations. Diamond powders may be used for roughing but finer powders are used for finishing.

After the shaping, the lens has to be centred and then reduced to its final diameter. Centring and shaping are sometimes simultaneous Some of the processes are automated but a larger labour content is still involved in the total product cost for lenses. The small diameter endoscope lenses pose problems of mounting for shaping.

Appendix Three

BATTERIES

Batteries are used as an alternative to a mains supply for some quartz halogen and incandescent light sources when safety and/or portability are important. They are also used to provide continuity of supply for some scanning units. The choice of battery is dependent on the following criteria: weight, life, rechargeability, discharge characteristics, load capability and environmental acceptability.

There are five main types of dry battery:

1   Zinc carbon
2   Mangenese alkaline.
3   Mercury.
4   Sealed lead acid.
5   Nickel cadmium.

1   Zinc carbon batteries are disposable and have a liquid electrolyte, although they are, for all practical purposes, dry. They have the zinc can as their negative electrode and the positive is the central carbon rod with its surrounding carbon powder. The electrolyte is ammonium chloride. For light sources that screw onto fibre-optic viewing instruments, their 1.5-V output is used in series. Their discharge curves are almost linear. As the chemical reaction proceeds, hydrogen bubbles are released. These insulate the carbon and progressively limit the cell's action. High-power versions of these batteries are available.

Layered batteries closely resemble zinc carbon batteries and are assembled in series by the manufacturer to yield higher voltages. They are usually rectangular in shape as opposed to cylindrical.

2   *Manganese alkaline* batteries are cylindrical in shape and disposable. They have better load capabilities than their zinc carbon counterparts. The positive electrode is compressed manganese dioxide, the can is the negative electrode and potassium hydroxide is the electrolyte. The manganese dioxide acts as a depolariser which resists the impairment of the battery's action by hydrogen bubbles.

3   *Mercury cells* are very light in weight and disposable. Zinc is again the negative and the positive electrode is compressed graphite and mercuric oxide. Their advantage is in their discharge characteristic. The voltage remains almost constant for the life of the battery.

4   *Sealed lead acid accumulators*   The type of accumulator used with fibre-optic lamps is rated for 6 V and, unlike a car battery, can be carried strapped around the waist. For hazardous environments, the accumulators can be sealed. In this type of battery the plates are grids of lead/antimony alloy. These support the active materials which are in porous form. The negative plate supports lead and the positive plate lead oxide (PbO_2). During discharge lead oxide is converted to lead sulphate ($PbSO_4$) by the action of the sulphuric acid electrolyte.

Sealed lead acid batteries are cylindrical in shape, rechargeable and marginally heavier, but lower in cost, than nickel cadmium batteries, but they have lower capacity. Their discharge charateristics resemble those of the mercury cell. The electrodes are thin strips of spirally wound pure lead.

5   *Nickel cadmium batteries,* which are used in scanning devices, are small, rechargeable, lightweight and durable. Nickel oxide acts as the positive electrode and cadmium as the negative. Sintered electrodes are used for applications requiring a high rate of discharge, e.g. 4 A-hr, extreme ambient temperatures and a fast charge. Potassium hydroxide is the electrolyte and the nominal voltage is 1.5 V. They can be stored indefinitely in any state of charge (unlike lead acid batteries). They incorporate a vent for discharging gases that might be produced in fault conditions; these gases may be unacceptable in hazardous environments.